MARCUS DU SAUTOY

Finding Moonshine

HARPER PERENNIAL
London, New York, Toronto, Sydney and New Delhi

KU-164-357

Harper Perennial
An imprint of HarperCollins*Publishers*
77–85 Fulham Palace Road
Hammersmith
London W6 8JB

www.harperperennial.co.uk
Visit our authors' blog at www.fifthestate.co.uk
Love this book? www.bookarmy.com

This Harper Perennial edition published 2009
2

First published in Great Britian by Fourth Estate in 2008

Copyright © Marcus du Sautoy 2008

PS Section copyright © Roger Tagholm 2009, except 'Einstein, Plato . . . and you?'
by Marcus du Sautoy © Marcus du Sautoy 2008, reproduced by kind permission
of the Telegraph Media Group.

PS™ is a trademark of HarperCollins*Publishers* Ltd

Marcus du Sautoy asserts the moral right to be identified as the author of this work

A catalogue record for this book is available from the British Library

ISBN 978–0–00–721462–4

Set in Postscript Linotype Minion by Rowland Phototypesetting Ltd

Printed and bound in Great Britain by Clays Ltd, St Ives plc

Mixed Sources

Product group from well-managed
forests and other controlled sources
www.fsc.org Cert no. SW-COC-1806
© 1996 Forest Stewardship Council

FSC

FSC is a non-profit international organisation established to promote the
responsible management of the world's forests. Products carrying the FSC
label are independently certified to assure consumers that they come
from forests that are managed to meet the social, economic and
ecological needs of present and future generations.

Find out more about HarperCollins and the environment at
www.harpercollins.co.uk/green

All rights reserved. No part of this publication may be reproduced, stored in a
retrieval system, or transmitted, in any form or by any means, electronic,
mechanical, photocopying, recording or otherwise, without the prior
permission of the publishers.

This book is sold subject to the condition that it shall not, by way of trade
or otherwise, be lent, re-sold, hired out or otherwise circulated without
the publisher's prior consent in any form of binding or cover other than
that in which it is published and without a similar condition including this
condition being imposed on the subsequent purchaser.

For Tomer, Magaly and Ina
and my mathematical children.

Contents

August: Endings and Beginnings

The universe is built on a plan the profound symmetry of which
is somehow present in the inner structure of our intellect.

PAUL VALÉRY

Midday, 26 August, the Sinai Desert

It's my 40th birthday. It's 40 degrees. I'm covered in factor 40 sun
cream, hiding in the shade of a reed shack on one side of the Red Sea.
Saudi Arabia shimmers across the blue water. Out to sea, waves break
where the coral cliff descends to the sea floor. The mountains of Sinai
tower behind me.

I'm not usually terribly bothered by birthdays, but for a mathematician 40 is significant – not because of arcane and fantastical
numerology, but because there is a generally held belief that by 40 you
have done your best work. Mathematics, it is said, is a young man's
game. Now that I have spent 40 years roaming the mathematical
gardens, is Sinai an ominous place to find myself, in a barren desert
where an exiled nation wandered for 40 years? The Fields Medal, which
is mathematics' highest accolade, is awarded only to mathematicians
under the age of 40. They are distributed every four years. This time
next year, the latest batch will be announced in Madrid, but I am now
too old to aspire to be on the list.

As a child, I hadn't wanted to be a mathematician at all. I'd decided
at an early age that I was going to study languages at university. This,
I realized, was the secret to fulfilling my ultimate dream: to become a
spy. My mum had been in the Foreign Office before she got married.

The Diplomatic Corps in the 1960s didn't believe that motherhood was compatible with being a diplomat, so she left the Service. But according to her, they'd let her keep the little black gun that every member of the Foreign Office was required to carry. 'You never know when you might be recalled for some secret assignment overseas,' she said, enigmatically. The gun, she claimed, was hidden somewhere in our house.

I searched high and low for the weapon, but they'd obviously been very thorough when they taught my mum the art of concealment. The only way to get my own gun was to join the Foreign Office myself and become a spy. And if I was going to look useful, I'd better be able to speak Russian.

At school I signed up for every language possible: French, German and Latin. The BBC started running a Russian course on television. My French teacher, Mr Brown, tried to help me with it. But I could never get my mouth around saying 'hello' – *zdravstvuyte* – and even after eight weeks of following the course I still couldn't pronounce it. I began to despair. I was also becoming increasingly frustrated by the fact that there was no logic behind why certain foreign verbs behaved the way they did, and why certain nouns were masculine or feminine. Latin did hold out some hope, its strict grammar appealing to my emerging desire for things which were part of some consistent, logical scheme and not just apparently random associations. Or perhaps it was because the teacher always used my name for second-declension nouns: Marcus, Marce, Marcum, . . .

One day, when I was 12, my mathematics teacher pointed at me during a class and said, 'du Sautoy, see me at the end of the lesson.' I thought I must be in trouble. I followed him outside, and when we reached the back of the maths block he took a cigar from his pocket. He explained that this is where he came to smoke at break-time. The other teachers didn't like the smoke in the common room. He lit the cigar slowly and said to me, 'I think you should find out what mathematics is really about.'

I don't quite know even now why he singled me out from all the others in the class for this revelation. I was far from being a maths prodigy, and lots of my friends seemed just as good at the subject. But something obviously made Mr Bailson think that I might have an appetite for finding out what lay beyond the arithmetic of the classroom.

He told me that I should read Martin Gardener's column in *Scientific American*. He gave me the names of a couple of books which he thought I might enjoy, including one called *The Language of Mathematics*, by Frank Land. The simple fact of a teacher taking a personal interest in me was enough to spur me on to investigate what it was that he found so intriguing about the subject.

That weekend my dad and I took a trip up to Oxford, the nearest academic city to our home. A little shopfront on The Broad bore the name Blackwell's. It didn't look terribly promising, but someone had told my dad that this was the Mecca of academic bookshops. Entering the shop you realized why. Like Doctor Who's Tardis, the shop was huge once you had entered the tiny front door. Mathematics books, we were told, were down in the Norrington Room, as the basement was known.

As we went downstairs a vast cavernous room opened up before us, stuffed full of what looked to me like every possible science book that could ever have been published. It was an Aladdin's cave of science books. We found the shelves dedicated to mathematics. While my dad searched for the books my teacher had recommended, I started pulling books off the shelves and peering inside. For some reason there seemed to be a high concentration of yellow books. But it was what I found within the yellow covers that grabbed my attention. The contents looked extraordinary. I recognized strings of Greek letters from my brief foray into learning Greek. There were storms of tiny little numbers and letters adorning x's and y's. On every page there were words in bold like **Lemma** and **Proof**.

It was completely meaningless to me. There were a few students leaning against the bookshelves who seemed to be reading the books as though they were novels. Clearly, they understood this language. It was simply code for something. From that moment I decided that I was going to learn how to decode these mathematical hieroglyphics. As we were paying at the till, I saw a table full of yellow paperbacks. 'They're mathematical journals,' explained the shop assistant. 'The publishers are offering free copies to entice academics to take out a subscription.'

I picked up a copy of something called *Inventiones Mathematica* and put it in the bag with the books we'd just bought. Here was my challenge. Could I decode the mathematical inventions in this yellow book? Some of the articles were in German, one was in French and the rest were in English. But it was the mathematical language that I was now determined to crack. What did 'Hilbert space' and 'isomorphism problem' mean? What message was hidden in these lines of sigmas and deltas and symbols that I couldn't even name?

When I got home I started looking at the books we'd bought. *The Language of Mathematics* particularly intrigued me. Before our expedition to Oxford, I'd never thought of mathematics as a language. At school it seemed to be just numbers that you could multiply or divide, add or subtract, with varying degrees of difficulty. But as I looked through this book I could see why my teacher had told me to 'find out what maths is really about'.

In this book there was no long division to lots of decimal places or anything like that. Instead there were, for example, important number sequences like the Fibonacci numbers. Apparently, the book said, these numbers explain how flowers and shells grow. You get any number in the sequence by adding the two previous numbers together. The sequence starts 1, 1, 2, 3, 5, 8, 13, 21, ... The book explained how these numbers are like a code that tells a shell what to do next as it grows. A tiny snail starts off with a little 1×1 square house. Then, each time it outgrows its shell, it adds another room to the house. But

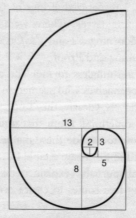

Fig. 1 How the snail uses the Fibonacci numbers to grow its shell.

since it doesn't have much to go on, it simply adds a room whose dimensions are the sum of the dimensions of the two previous rooms. The result of this growth is a spiral (Figure 1). It was beautiful and simple. These numbers are fundamental, said the book, to the way nature grows things.

Other pages depicted interesting three-dimensional objects that I'd never seen before, built from pentagons and triangles. One was called an icosahedron and had 20 triangular faces (Figure 2). Apparently, if you took one of these objects (what the book called polyhedra) and counted the number of faces and points (what the book called vertices), and then subtracted the number of edges, you always got 2. For example, a cube has 6 faces, 8 vertices and 12 edges: $6 + 8 - 12 = 2$. The book claimed that this trick would work for any polyhedron. That seemed like a bit of magic. I tried it on the one made out of 20 triangles.

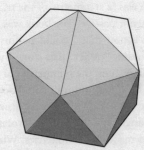

Fig. 2 The icosahedron with its 20 triangular faces.

The trouble was that it was quite hard to envisage the whole object clearly enough to count everything. Even if I built one from card, keeping track of all those edges seemed a bit daunting. But then my dad showed me a short cut. 'How many triangles are there?' Well, the book said that there were 20. 'So that's 60 edges on 20 triangles, but each edge is shared by two triangles. That makes 30 edges.' Now, that really was magic. Without looking at the icosahedron, you could work out how many edges it had. The same trick worked for the vertices. Again, 20 triangles have 60 vertices. But this time I could see from the picture that every vertex was shared by five triangles. So the icosahedron had 20 faces, 12 vertices and 30 edges. And sure enough, $20 + 12 - 30 = 2$. But why did the formula work whatever polyhedron you took?

In another book there was a whole section on the symmetry of objects like these polyhedra made out of triangles. I had a vague idea of what 'symmetry' meant. I knew that I was symmetrical, at least on the outside. Whatever I had on the left side of my body, there was a mirror image of it on the right side. But a triangle, it seemed, had much more symmetry than just the simple mirror symmetry. You could spin it round as well, and the triangle still looked the same. I began to realize that I wasn't actually sure what it meant to say that something was symmetrical.

The book stated that the equilateral triangle had six symmetries. As I read on, I began to see that the triangle's symmetry was captured by the things I could do to it that would leave it looking the same. I traced an outline around a triangular piece of card and then counted the number of ways I could pick the triangle up and put it down so that it fitted back exactly inside its outline on the paper. Each of these moves, the book said, was 'a symmetry' of the triangle. So a symmetry was something active, not passive. The book was pushing me to think of a symmetry as an action that I could perform on the triangle to replace it inside its outline, rather than some innate property of the triangle itself. I started to count the symmetries of the triangle, thinking of them as the various different things I could do to it. I could flip the triangle over in three ways. Each time two corners swapped places. I could also spin the triangle by a third of a full rotation, either clockwise or anticlockwise. That made five symmetries. What was the sixth?

I searched desperately for what I'd missed. I tried combining actions to see whether I could get a new one. After all, performing two of these moves one after the other was effectively the same as making a single move. If a symmetry was a move that put the triangle back inside its outline, then perhaps I would get a new move or a new symmetry. What if I flipped the triangle then turned it? No, that was just like one of the other flips. What about flipping, rotating and then flipping back again? No, that just created the spin in the other direction, which I'd counted already. I'd got five things, but whatever combination I took of these moves I couldn't get anything new. So I went back to the book.

What I found was that they'd included as a symmetry just leaving the triangle where it was. Curious . . . But I soon saw that if symmetry meant anything you could do to the triangle that kept it inside its outline, then not touching it at all – or, equivalently, picking it up and putting it back in exactly the same place – was also an action that had to be included.

I liked this idea of symmetry. The symmetries of an object seemed to be a bit like all the magic trick moves. The mathematician shows you the triangle, then tells you to turn away. While you are not looking, the mathematician does something to the triangle. But when you turn back it looks exactly as it did before. You could think of the total symmetry of an object as all the moves that the mathematician could make to trick you into thinking that he hadn't touched it at all.

I tried out this new magic on some other shapes. Here was an interesting one, looking like a six-pointed starfish (Figure 3). I couldn't flip it over without making it look different: it seemed to be spinning in one direction, which destroyed its reflectional mirror symmetry. But I could still spin it. With its six tentacles, there were five spins I could do, together with just leaving it where it was. Six symmetries. The same number as the triangle.

Fig. 3 A six-pointed starfish with no reflectional symmetry.

Each object had the same number of symmetries. But the book talked about a language that could articulate and give meaning to the statement 'These two objects have different symmetries.' It would reveal why these objects represented two different species in the world of symmetry. This language could also expose, the book promised, when two objects that looked physically different actually had the same symmetries. This was the journey I was about to embark on: to discover what symmetry really is.

As I read on, the shapes and pictures gave way to symbols. Here

was the language that the title of the other book was referring to. There
seemed to be a way to translate the pictures into a language. I came
across some of the symbols that I'd seen in the yellow journal I'd
picked up. Everything was starting to get rather abstract, but it seemed
that this language was trying to capture the discovery I'd made when
playing with the six symmetries of the triangle. If you took two symmetries, or magic trick moves, and did them one after the other, for
example a reflection followed by a rotation, it gave you a third symmetry. The language describing these interactions had a name: group
theory.

This language provided an insight into why the six symmetries of
the six-pointed starfish were different to the six symmetries of the
triangle. A symmetry was one of these magic trick moves, so I could
perform two symmetries of an object one after the other to get a third
symmetry. The group of symmetries of the starfish interact with one
another very differently to the interaction between the group of symmetries of the triangle. It was the interactions among the group of
symmetries of an object that distinguished the group of symmetries of
the triangle from the group of symmetries of the six-pointed starfish.

In the starfish, for example, one rotation followed by another gave
me a third rotation. But it didn't matter in what order I made the
two rotations. For example, spinning the starfish 180° clockwise then
anticlockwise 60° left the starfish in the same position as first doing
the 60° anticlockwise spin and then the 180° clockwise spin. In contrast,
if I took two symmetries of the triangle and combined the two magic
trick moves corresponding to these symmetries, it made a big difference what order I did them in. A mirror symmetry move followed by
a rotation was not the same as the rotation followed by the mirror
symmetry move. The language of my book had translated the pictures
into the sentence $M \cdot R \neq R \cdot M$, where M was the mirror symmetry move
and R the rotation (Figure 4). The physical world of symmetry could
be translated into an abstract algebraic language.

As my school years progressed, I came to see what my maths teacher
had done. The arithmetic of the classroom is a bit like scales and
arpeggios for a musician. My teacher had played me some of the
exciting music that was waiting for me out there if I could master the
technical part of the subject. I certainly didn't understand everything
I read, but I did now want to know more.

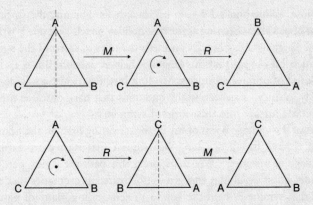

Fig. 4 A mirror symmetry followed by a rotation is different from a rotation followed by a mirror symmetry.

Most budding musicians would abandon their instruments if all they were allowed to play and listen to were scales and arpeggios. A child starting out on an instrument will have no idea how Bach composed the *Goldberg Variations* or how to improvise a blues lick, yet they can still get a kick out of hearing someone else do it. Books such as *The Language of Mathematics* made me realize that you could do the same with maths. I didn't have a clue what 'a group' really was, but I grasped that it was part of a secret language that could be used to unlock the science of symmetry.

This was the language I would try to learn. It might not get me into the Foreign Office, and I might have to give up the dream of being a spy, but here was a secret code that looked as intriguing as anything the world of espionage might throw up. And unlike Russian or German, this language of mathematics seemed to be a perfect idealized language in which everything made sense and there were no irregular verbs or nonsensical exceptions.

Of all the things I had seen in those books, it was group theory – the language of symmetry – that intrigued me most. It seemed to take a world that was full of pictures and turn it into words. The dangerous ambiguities that plague the visual world, with its plethora of optical illusions and mirages, were made transparent by the power of this new grammar.

I've been sitting on the beach in the shade of our shack reading one of those yellow books I'd seen in Blackwell's. For me, the stories in those books are as exciting as the best holiday novel. This one is written in the language of symmetry and tells the tales of some of the strange symmetrical objects that this language helped unleash. But it also is a book full of unfinished stories. My 40th birthday is just a staging post on my journey to answering the questions that have obsessed me as I journeyed further into this world of symmetry.

From the vantage point of my birthday, sitting here on the beach in Sinai, I have travelled a long way since I first started to learn the language of symmetry. My steps along this path are a tiny part of a grander quest which has engaged mathematicians ever since they realized that symmetry held the key to understanding many of nature's intimate secrets.

Nature's language

The sun is setting behind the mountains of Sinai, and the tide is receding across the coral shelf that runs parallel to the coastline. It is time for white men and crustaceans to emerge from the shade. A bit of exercise might help sort out the mess in my head. There are two Israeli guys up ahead who are staying in the Bedouin camp. For them, Sinai is a welcome escape from guard duty in Gaza. Their backs are scorched from snorkelling too long in the Sinai sun. They're pointing excitedly into the water, intrigued by something they've found on the surface of the coral. When I look down, I suddenly notice the coral surface is covered with one of nature's most remarkable symmetrical animals.

There in the water is a real starfish like the picture I'd played with as a child. I'm not sure if I've ever seen a live starfish before. This one has the classic five tentacles that most people associate with starfish, but it is not as rigid as the cartoon-style crustaceans I'm used to seeing. Apparently some starfish, not content with the simple five-pointed pentacle, have gone for even showier displays of symmetry. The sunflower starfish starts out life with five legs, but during its eight-year life span it can grow as many as 24 legs. Being able to generate a shape which looks exactly the same in 24 different directions is some feat of biological engineering.

Why, though, is symmetry so pervasive in nature? It is not just a matter of aesthetics. Just as it is for me and mathematics, symmetry in nature is about language. It provides a way for animals and plants to convey a multitude of messages, from genetic superiority to nutritional information. Symmetry is often a sign of meaning, and can therefore be interpreted as a very basic, almost primeval form of communication. For an insect such as the bee, symmetry is fundamental to survival.

The eyesight of the bee is extremely limited. As it flies round negotiating the world, its brain receives images that are as distorted as if we were looking at the world through a thick sheet of glass. The bee can't judge distances, so it continually crashes into things. The bee suffers a form of colour-blindness. The background green of the garden appears grey; red stands out more clearly as a blackness against the grey. But even through this thick-rimmed pair of glasses, there is one thing that burns strongly in the eyes of the bee: symmetry.

The honeybee likes the pentagonal symmetry of honeysuckle, the hexagonal shape of the clematis, and the highly radial symmetry of the daisy or sunflower. The bumblebee prefers mirror symmetry, such as the symmetry of the orchid, pea or foxglove. The eyesight of bees has evolved sufficiently for them to pick out these significant shapes. For in symmetry there is sustenance. The bees that are drawn to shapes with pattern are the insects that will not go hungry. For the bee, survival of the fittest means becoming an expert at symmetry. The bee that could not read the signs and signals of sustenance was left buzzing randomly round the garden, unable to keep up with its superior competitors who could spot the patterns.

Because the plant is equally dependent on attracting the bee to its flower for pollination and prolonging its genetic heritage, it too has played its part in this natural dialogue. The flower that can achieve perfect symmetry attracts more bees and survives longer in the evolutionary battle. Symmetry is the language used by the flower and bee to communicate with each other. For the flower, the hexagon or the pentagon is like a billboard shouting out 'Visit me!' For the bee, encoded in the symmetrical shape is the message that 'Here is food!' Symmetry denotes something special, something with meaning. Against the static white noise that makes up most of the bee's visual world, the six perfect petals of the clematis stand out like a musical phrase full of harmony.

As nature's garden evolved, so too did the variety of shapes and colours exploited by the plant world. After millions of years of spring following winter to produce another year of geometric evolution, the garden is now a plethora of patterns trumpeting their greetings and promises of sweet sustenance.

But symmetry is not an easy thing to achieve. A plant has to work hard and be able to divert important natural resources to achieve the balance and beauty of the orchid or the sunflower. Beauty of form is an extravagance. That is why only the fittest and healthiest individual plants have enough energy to spare to create a shape with balance. The superiority of the symmetrical flower is reflected in a greater production of nectar, and that nectar has a higher sugar content. Symmetry tastes sweet.

The flower or animal with symmetry is sending out a very clear signal of its genetic superiority over its neighbours. That is why the animal world is populated by shapes that strive for perfect balance. Humans and animals are genetically programmed to look upon these shapes as beautiful – we are attracted to those animals whose genetic make-up is so superior that they can use energy to make symmetry.

Humans and animals alike will choose a face that has perfect left–right mirror symmetry over an unsymmetrical face. Most of the animals in the natural world favour such bilateral mirror symmetry. A line down the middle separates the shape into two different halves. But although they are different, there is a perfect correspondence which matches one half to the other. At least externally. The asymmetry of our internal organs is still something of a mystery and only goes to reinforce the wonder at how symmetrical the exterior is.

Studies indicate that the more symmetrical among us are more likely to start having sex at an earlier age. Even the smell men emit seems to be more appealing to women when the male has more symmetry. In one study, sweaty T-shirts that had been worn by men were offered to a selection of women, and those who were ovulating were drawn to the tee-shirts worn by the men with the most symmetrical bodies. It seems, though, that men are not programmed to pick up the scent of a symmetrical woman.

Animal rights activists have used symmetry as evidence of cruelty to animals. Battery farm eggs are likely to be far less symmetrical than free-range eggs: battery hens are suffering trauma and wasting energy that could have been used to realize perfection. Unlike the tortured artist thriving in adversity to create great art, the hen needs comfort and luxury to produce perfect symmetry.

Animals have also been drawn to mirror symmetry because of the superior motor skills it offers. Symmetry is often associated with the idea of a shape being in perfect balance – one half with another. Nearly all motor abilities are reliant on symmetry to propel them in the most efficient manner. It is the most symmetrical two- and four-legged members of a species who can move the fastest. The food goes to the animal with the most symmetry because it's going to get to the dinner table first. Similarly, the prey who can run fastest stands the best chance of avoiding becoming dinner. So natural selection favours the form that creates the fastest animal – and balance in motion is intimately tied up with symmetry of form. The animal with one leg much longer than the others is going to run round in circles and won't survive the fierce pace of natural selection.

But symmetry isn't just a genetic language for declaring to potential mates how good one's DNA is. Back in the hive, away from the search for symmetrical flowers and nectar, symmetry also pervades the bee's home life. As the young bees gorge themselves on the honey that has been collected, they secrete small slivers of wax. The temperature of the hive is maintained at 35°C by the concentration of bees, which makes the wax malleable enough to be shaped by the worker bees, who collect the wax secretions and mould the cells in which the honey will be stored. The hexagonal lattice that the bees use to store their honey exploits another facet of symmetry. Not only is it a harbinger of meaning and language, but also symmetry is nature's way of being efficient and economical. For the bee, the lattice of hexagons allows the colony to pack the most honey into the greatest space without wasting too much wax on building its walls.

Although bees have known for ages that hexagons are the most efficient shape for building a honey store, it is only very recently that mathematicians have fully explained the Honeycomb Conjecture: from the infinite choice of different structures that the bees could have built, it is hexagons that use the least wax to create the most cells.

Although symmetry is genetically hard to achieve, many natural phenomena will gravitate towards symmetry as the most stable and efficient state. The inanimate world is full of examples of the drive for symmetry of form. When a soap bubble forms it tries to assume the shape of a perfect sphere, the three-dimensional shape with the most symmetry. However much you rotate or reflect a sphere, its shape still looks the same. But for the soap film it is the efficiency of the shape of the sphere that appeals. The energy in the soap film is directly proportional to the surface area of the bubble. The sphere is the shape with the smallest surface area that can contain a given volume of air, and hence it is the shape that uses the least energy. Like a stone rolling down a mountain to the point of lowest energy in the valley below, the symmetrical sphere represents the optimal shape for the soap film.

The raindrop as it falls through the sky is not in fact the tear shape that artists often paint – that's just an artistic convention to give a sense of rain in motion. The true picture of a drop of water falling from the sky is a perfect sphere. Lead shot manufacturers have exploited this fact since the eighteenth century: molten lead is dropped from a great height into buckets of cold water to make perfectly spherical balls.

Scientists have discovered mysterious symmetries hiding at the heart of many parts of the natural world – fundamental physics, biology and chemistry all depend on a complex variety of symmetrical objects. The snowflake and the deadly HIV virus both exploit symmetry. In the chemical world, a diamond gets its strength from its highly symmetrical arrangement of carbon atoms. In physics, scientists established the connection between electricity and magnetism by discovering how these parts are simply two different sides of a common symmetrical phenomenon. New fundamental particles have been predicted thanks to spinning through the symmetries of strange shapes. The different symmetries hint at the existence of new particles which are mirroring particles we already understand.

For as long as humans have been communicating with each other, symmetry has remained a central idea in the lexicon. Repeating patterns is key to how a baby first learns language. Symmetry continues to inform the way we craft words in songs and poetry. From the first cave paintings to modern art, from primitive drumbeats to contemporary music, artists have continually pushed symmetry to the extremes. As with the humble bee, symmetry has provided manufacturers with efficient ways to create and build, from the Arab carpet weavers to the engineers who have managed to encode more and more data onto smaller and smaller electronic devices. Symmetry is behind every step in our evolutionary development.

The word 'symmetry' conjures to mind objects which are well balanced, with perfect proportions. Such objects capture a sense of beauty and form. The human mind is constantly drawn to anything that embodies some aspect of symmetry. Our brain seems programmed to notice and search for order and structure. Artwork, architecture and music from ancient times to the present day play on the idea of things which mirror each other in interesting ways. Symmetry is about connections between different parts of the same object. It sets up a natural internal dialogue in the shape.

I can't step over the starfish in the sea without spinning the pentacle in my head. I can't ignore the strange pattern that adorns my swimming trunks. Even footsteps in the sand get me thinking about a problem that I can't stop exploring once it's occurred to me. How many different ways can I mark out shapes in the sand as I make my way along the beach? My simple footsteps are something called a glide reflection – each step is got by reflecting the previous footstep then gliding it across the sand. Now I hop along the beach kangaroo-fashion, and my two feet create a pattern with simple reflection. When I spin in the air and land facing the other way, I get a pattern with two lines of reflectional symmetry. In all, I manage to make seven different symmetries in the sand. The Bedouin fishermen who are catching our dinner are laughing at me as I jump and hop around in my exploration of symmetry in the sand.

The symmetry seekers

Mathematics is sometimes called the quest for patterns. Jumping about in the sand, I found I could make seven different types of pattern with my footprints. But is it possible to classify all the possible patterns that could be found in nature? Is there a limit to what patterns we might find? Could we even make a list of all these possible symmetries? For the mathematician, the pattern searcher, understanding symmetry is one of the principal themes in the quest to chart the mathematical world.

For several millennia, mathematicians have been gradually

accumulating symmetrical shapes as they explored further and further
afield. But symmetry is a slippery concept. What exactly is it? When
do two objects have the same symmetries and when are they different?
It took a stunning breakthrough during the revolutionary fervour of
nineteenth-century Paris for a new language to emerge that could
capture the true meaning of the word. As I'd learnt from the book my
teacher had recommended, it was called group theory. This new language became the seed for a mathematical revolution which would
match in its implications the political upheaval then taking place on
the streets of Paris. Suddenly, mathematics had the tools to build ships
to set sail for the very limits of the world of symmetry.

One of the most important discoveries revealed by this new
nineteenth-century language of group theory was that behind symmetry lay a concept of prime building blocks. The Ancient Greeks
knew that every number can be divided into prime numbers – indivisible numbers – and that these numbers were the building blocks of all
other numbers. The nineteenth-century language for symmetry threw
up the far subtler fact that, just like the division of numbers, every
symmetrical object could also be divided into certain smaller objects
whose collection of symmetries were indivisible. For example, the
rotations of a 15-sided figure could be built from the rotations of a
pentagon and the rotations of a triangle. But the group of rotations of
these 'prime-sided' figures could not be divided up into smaller groups
of symmetries. The group of symmetries of the pentagon was an
indivisible group of symmetries. The crucial thing about these indivisible groups of symmetries was the fact that they were the building
blocks from which all symmetrical objects could be built. Just as the
prime number 5 is a building block of larger numbers, the pentagon
was one of the building blocks in the world of symmetry.

It took mathematicians a long time to fully grasp the idea of what
made a symmetrical object indivisible. But when they did, they saw
the prospect of producing a 'periodic table' of symmetry consisting of
all the different possible indivisible symmetrical objects, in the same
way that chemistry's periodic table collects together the chemically
indivisible elements from which all other substances are made. Such a
table would list all the building blocks out of which all possible symmetrical objects can be constructed. Prime numbers are the key to the
first objects to be included in the periodic table of symmetry: the

rotational symmetries of a prime-sided polygon or coin. But in the world of symmetry there turned out to be other, stranger objects whose symmetries were indivisible. One of the first of these more exotic building blocks of symmetry was the rotational symmetries of the icosahedron with its 20 triangular faces. The mathematicians of the nineteenth century discovered that the icosahedron was an object whose symmetries could not be reduced to smaller objects.

Ever since the Ancient Greeks discovered the icosahedron thousands of years ago, mathematicians have been marvelling at and exploring the world of symmetry. But this new window opened up by group theory offered the prospect of mastering and classifying this world. If you knew the building blocks of symmetry, you could become symmetry's architect. The mathematicians of the nineteenth and twentieth centuries unearthed and added more and more indivisible symmetrical objects to this mathematical periodic table. But the list just kept on growing, and they began to wonder whether a list of all possible indivisible symmetrical objects could ever be completed.

Then, in the 1970s, along came a band of mathematical explorers whose skills, determination and sheer persistence were equal to the task of navigating the limits of this complex world. The explorers divided into two distinct teams. One specialized in finding more and more exotic mathematical objects whose symmetries were indivisible. Like pirates hunting for treasure, this was the fun team to be in, looking out for new building blocks of symmetry. But the stakes were high. While a few of them carved their names into the annals of symmetry with their discoveries, many searched in vain and returned empty-handed. Luck as much as judgement was an important factor in whether there was treasure at the end of any particular rainbow.

In contrast to the swashbuckling of this first team, the second one consisted of a more disciplined fighting force. This well organized troop worked from the other end, exploiting the limitations of symmetry. They soberly assessed each twist and turn, explaining why there were no new indivisible symmetries that could possibly exist if you set off in certain directions.

The first team consisted of a ramshackle collection of mathematical mavericks. One of the most colourful was John Horton Conway, currently professor at the University of Princeton. His mathematical and personal charisma have given him almost cult status. Conway's performances when he presents the spoils of his mathematical raids are almost magical in quality. He weaves together what at first sight look like mathematical curios or tricks, but by the end of the lecture has arrived at answers to very fundamental questions of mathematics. Each revelation of a fundamental insight is preceded by his characteristic laugh, as if he too is surprised at where he has arrived. At the same time he has reduced a room of serious academics to playful children. They rush up at the end of the lecture to play with the mathematical toys he produces from a suitcase of tricks that he often carries with him.

At the helm of the second team was Daniel Gorenstein. During the 1960s, hundreds of mathematicians around the world turned their attention to understanding the limits of the world of symmetry. Their efforts were focused more on showing what was not possible. In 1972 Gorenstein decided that a coordinated attack combining everyone's individual skills was needed. Without his stewardship, mathematicians might still have been wandering the globe unaware of each other's progress. Advances were sometimes painstaking and treacherous as they battled their way through complex and lengthy proofs, some extending to thousands of pages of logical argument. Gorenstein often referred to those decades of exploration as the Thirty Years War.

While the first team of explorers plundered new territories, the second team systematically surveyed what was and was not possible. Would the second team ever be able to show the first team that there was no longer anywhere new to explore? Or would the world of symmetry turn out not to be a closed globe but an infinite expanse that would see these two teams journeying for ever, destined never to close the loop? Might there always be uncharted waters? Many in the first team hoped that the journey would go on for ever, revealing ever more exotic symmetries. But the second team hankered after closure and complete knowledge.

Towards the end of the 1970s, mathematicians realized that the two teams were finally closing in on each other. A complete taxonomy of symmetry was in sight – a periodic table containing all the building blocks of symmetry was emerging. Most mathematicians were thrilled at the prospect of a proof that the symmetry seekers had found all the building blocks. But not all were happy. The pirate captain, John Conway, was asked whether he was optimistic or pessimistic about such a prospect. Rather enigmatically he replied, 'A pessimist, but still hopeful . . . I was delighted to find that the answer was misinterpreted in exactly the way I had maliciously desired!' For a treasure seeker like Conway, these symmetrical objects were 'beautiful things, and I'd like to see more of them, but I am reluctantly coming round to the view that there are likely to be no more to be seen'.

In contrast, Gorenstein and his military cohort found optimism in finally seeing an end to the exploration with the cessation of the Thirty Years War. By the beginning of the 1980s two more indivisible symmetrical building blocks were added to the list, but at that point they could see the other team on the horizon. As the teams approached each other, people began to realize: that's it. No more surprises out there. The world of symmetry had been circumnavigated. Word started to spread in 1980 that the search was over, the classification complete. But it was a strange ending to such an epic journey. There was no climactic moment when a mathematician put down the chalk and the audience rose to their feet to applaud the great achievement. There were no press conferences to announce that finally the proof had been finished. No one is even too clear who had actually finished it. Some still question whether it truly has been completed.

It was not something that made the news outside the mathematical community. At the time I was in the sixth form. My bedroom wall at home was covered not with posters of bands or football stars, but newspaper cuttings about mathematics. I would trawl through the papers for exciting breakthroughs to stick up on my wall. I had recently been looking through the numerous articles I'd cut out, but not one made any mention of this phenomenal achievement. Intriguingly, I did discover that one of the cuttings I'd had next to my bed was a letter to the *Guardian* about a false proof of Fermat's Last Theorem that the newspaper had published a week earlier. The letter came from the mathematician who would later become my doctoral supervisor.

For the mathematical community, however, it was big news. It was a mammoth feat. For centuries no one had believed it possible: to write down a set of basic symmetrical objects that could be used to build all possible symmetrical objects. As mathematicians had gradually got to grips with what symmetry actually meant, they seemed to be gazing upon an endless world filled with a chaotic and infinitely varied range of symmetrical objects. That's why there was such a momentous sense of achievement in the mathematical community at what these mathematicians had done.

It was a mathematical proof like no other. Mathematicians are used to seeing the names of specific individuals attached to proofs of theorems: Andrew Wiles's proof of Fermat's Last Theorem, for example, or Grigori Perelman's proof of the Poincaré Conjecture. Mathematicians will hide away for years, working in isolation, determined to get their name on the theorem. But for the first time in mathematics, here was a proof which involved such a collective effort that it was impossible and meaningless to put a single name to it.

That said, the mathematicians who had discovered new islands of symmetry on the way had not been shy about planting a flag and getting their name on the map: the first, second, third and fourth Janko groups, the Harada–Norton group, Conway One, Two and Three. There were some bitter arguments about who had discovered certain groups of symmetries first and whose name the group should go by. But any attempt to name the proof of the classification would probably require at least a hundred different names to be attached to it.

Unlike any other proof, this one was so immense that it was unclear whether any one person could claim to have read all the ten thousand pages spread over five hundred different journals that completed its account. For many, such a proof went against the ethos of simplicity in mathematics, expressed in 1940 by the Cambridge mathematician G. H. Hardy: 'A mathematical proof should resemble a simple and clear-cut constellation, not a scattered cluster in the Milky Way.'

Although this wasn't an elegant one-line proof, it was as rich and varied as the wonders to be found in the Milky Way. Each new discovery by the symmetry seekers was greeted by mathematicians with as much excitement as the discovery of new moons and planets. Just as the Milky Way is an exotic treasure trove full of beautiful stars and nebulae, the proof, although vast and complex, is full of jewels that would have appealed to Hardy's sense of aesthetics. But the story of symmetry is different to the discoveries of astronomy. With crystal-clear logic, the new mathematical proof explained why all these symmetries should be out there and why we weren't going to find any more. There was no randomness in this arrangement. No other configuration would work.

Conway, the Long John Silver of mathematics, decided that an account should be published of the lands they had discovered on their voyage. Based in Cambridge, he was aided by Rob Curtis, Simon Norton, Richard Parker and Rob Wilson. Together they produced what is now known as 'the Atlas': mathematical charts documenting the topography of each new group of symmetries encountered.

Because so much of science depends on symmetry, this endeavour was not idle butterfly collecting. Huge swathes of mathematics, physics and chemistry can be explained in terms of the underlying symmetry of the structures under investigation. The Atlas of symmetry therefore became a Rosetta Stone for many scientists. Anyone faced with a question that reduced to understanding symmetry could now refer to this catalogue. Many mathematicians found that they could now prove their theorems simply by checking that the result is true for all the indivisible symmetrical building blocks in Conway & Co.'s Atlas. A famous number theorist at Harvard declared that if the library burnt down and he could rescue one book, it would be the Atlas of symmetry.

The charts in the Atlas are as fundamental to mathematics as the periodic table has been for chemists. For thousands of years, scientists had been striving to understand the basic constituents of matter itself. The Ancient Greeks had believed the building blocks to be earth, wind, fire and water. But twentieth-century chemistry settled on the periodic table originated by the Russian scientist Dmitri Mendeleev, which in its present-day form lists over a hundred chemical elements starting with hydrogen, helium and lithium. From the atoms of the elements in the table, one can build all the molecules in the known universe.

Now, two millennia after the Ancient Greeks had started to explore shapes with symmetry, mathematics had got its own periodic table. It lists the elements of the science of symmetry, the atoms from which all possible symmetries are built. But 'atlas' is a better word than 'table' for this huge red book which sits on many a mathematician's bookshelf. Inside are the contours, the towns and cities that make up every basic symmetrical territory.

Conway actually began compiling the Atlas years before anyone knew whether it would have a final page or whether it was destined to be an infinite volume. Once they knew that the journey was over, the Cambridge Five took their Atlas to their publisher to share the map with the scientific community. In 1985 this extraordinary document started running off the presses. The same year, I visited Cambridge as a spotty, twenty-year-old Oxford undergraduate hoping to begin my own personal journey into the world of symmetry.

Setting sail

After years of training at school and then university, practising my arithmetic scales and mathematical counterpoint, I was ready to start my own work. But I needed a mentor to help steer me in the right direction. My tutor at Oxford went through the list of group theorists at Cambridge and picked one out. 'Write to Simon Norton,' he said. We arranged to meet in the common room of the maths department at Cambridge.

I wasn't sure what Norton looked like, so faced with a common room full of mathematicians I felt a bit daunted. Like most mathematicians I am naturally quite shy. I'm not someone who likes to hold out my hand and introduce myself to people. I hate parties, and I'm terrified of the telephone. Mathematics had provided a safe haven full of things that didn't behave unexpectedly (or at least if they did, you knew that there was some perfectly logical explanation for their strange behaviour). What I loved about mathematics was that a proof spoke for itself: it didn't need you to present its credentials and persuade others of its validity. It was all there on the table.

No one seemed to be expecting me. Everyone seemed to have their head deep in something. Some were scribbling away animatedly on pads of paper, but most were engrossed in games of backgammon and go. I interrupted one of the groups to ask whether they could direct me towards Dr Norton.

A student pointed to the back of the common room: 'He's sitting over there.' I could see what looked like a tramp, with wild black hair sprouting out all over his head, trousers frayed at the turn-ups, wearing a shirt full of holes. He was surrounded by plastic bags which seemed to contain his worldly possessions. He looked like a scarecrow. 'Yeah, that's Simon.'

I went over and introduced myself. In a strange nasally voice with a hint of a nervous laugh, he said hello, but avoided any attempt of mine to shake his hand, recoiling as if I was about to assault him. Conversation was difficult. I'd met some pretty strange characters through my undergraduate studies, but no one like this. What seemed to excite him most was the route I'd taken from Oxford to Cambridge. He started producing bus and train timetables from his bags. Apparently there was an intriguing route I could have taken via Bletchley. Not that he needed the timetables: he seemed to know them all off by heart. He'd already planned my trip back.

While I sat there desperately trying to get some idea of where the future of group theory lay and getting instead a description of the nation's bus service, a large man bounded towards us and sat himself down next to Simon Norton. I wasn't quite sure who he was, but he seemed to think I should. He too had hair sprouting all over the place, this time ginger brown, and he grinned at me with a frighteningly wild glint in his eyes. It was deepest winter, but this man was happily sitting in sandals and a T-shirt with the decimal expansion of pi running across the whole stretch of his corpulent body. He looked like a slightly mad clown. As I was about to find out, this was John Conway, captain of the Cambridge ship.

I told him that I was interested in coming to Cambridge to do my PhD in group theory. 'What's your name ... with your initials?' 'Er ... Marcus du Sautoy, Marcus P. F. du Sautoy.' 'Drop the F. and the du, change the S of Sautoy to a Z and you can join us.' I hadn't a clue what he was talking about, and it obviously showed on my face. Had I failed some strange initiation rite? Or was this a strange puzzle I had to solve? Mathematicians can be quite cruel once they know how to do something and they enjoy seeing you squirm as you struggle to catch on. But I couldn't get this one.

He threw a big red book down in front of me. It landed with an impressive thwack on the square white table between us. On the front cover, *Atlas of Finite Groups*, and below the title five names:

J. H. Conway

R. T. Curtis

S. P. Norton

R. A. Parker

R. A. Wilson

'It's an atlas of symmetry. That's me at the top. Then, in order, those who joined the group.' Of course, now I got it. Each with two initials, each with a six-letter surname starting with a letter in the alphabet to denote the order they arrived in the group. I was only going to be let on board if my name was M. P. Zautoy. When you look inside, there is a sixth mathematician who is thanked for his computational assistance in preparing the book. But with a name like J. G. Thackray, he was never going to make it onto the front cover.

'When we first got it back from the printers, the typesetters had messed up the symmetry in our names. It was all misaligned. I insisted it go back to the printers and they do it all again.' The Atlas of symmetry would never have appeared were it not for people such as Conway who were so obsessed with symmetry as to insist on such details.

'I like symmetrical things. I've always loved gems and crystals and polyhedral shapes.' I could see this from the office he'd emerged from. It was crammed with symmetrical models of all shapes, sizes and colours, many hanging from the ceiling. They looked like an array of stellated candle holders from a Byzantine church. Conway's office was a shrine to symmetry.

'I've got a book with Escher's prints sitting on my piano,' he said. 'I try to ration myself to an Escher picture a day. Often I can't resist cheating and turning the page early, but I always insist on at least going out of the room first before I can turn the next page. One of my favourites is a picture of a tin box that Escher designed for a Dutch chocolate manufacturer [Figure 5]. It's an icosahedron made up of twenty triangles covered in starfish and shells. Escher was very clever. The starfish have all got a little twist on them so the five arms seem to spin anticlockwise. That means the shape doesn't have any reflective symmetry. Its only symmetries are the different rotations of the shape. Its symmetries are the first building block in the book.'

He flipped the book open to the first 'map' in the Atlas. At the top was its name, A_5, followed by a small table of numbers which provided the mathematical details of how to navigate the symmetries of this 'island'.

'When I'm interested in something I like to name it, list it, and then write a book about it. But if you want to make your name, then it's the penultimate entry in the Atlas you'll really want to understand.'

Fig. 5 Escher's icosahedral chocolate box.

Conway turned towards the back, to a page where the heading reads simply '*M*'. It sounded like the name of a spy, but he explained that *M* stands for Monster, a name he coined after the object was discovered. I'd heard some mention of this huge symmetrical object during my last undergraduate year. It certainly wasn't on the syllabus – it had been constructed for the first time in 1980. But I'd started going to some research seminars that year just to get a feel for what was out there beyond the weekly exercises dished out by our lecturers. I was quite shocked that despite having spent three years at university learning the language of mathematics, the seminars washed over me like a sea of meaningless words and symbols. It was obvious that I still had a long way to go. The Monster had figured in a number of seminars, but beyond an exotic sounding name I really had no idea what this object really was.

'It's got 808,017,424,794,512,875,886,459,904,961,710,757,005,754, 368,000,000,000 symmetries. That's why it's called the Monster.' I stared at him in amazement, not because the object had more symmetries than there are atoms in the sun, but because, without batting an eyelid, he could reel off the size of it. He could see that I was impressed. 'That's nothing. I could tell you all the digits on the back of my T-shirt too.' I looked at the shirt, which said '$\pi =$' followed by a huge string of digits. I could tell him the first six digits, 3.141 59, but that's as far as I can go. But Conway claimed he could recall the decimal expansion of π to thousands of decimal places. There weren't any obvious patterns in these numbers to help him generate them – not like the Fibonacci numbers, with their rule of adding two successive numbers to get the next in the sequence. But Conway has the sort of mind that can sniff out the least bit of structure to help him recall something so massively complex. And it's not an autistic mind, one that simply absorbs random information. Conway had taught himself these skills; his is an analytic mind that finds ways to perform such feats.

'Forget pi. It's these numbers that are really interesting,' he said, pointing to the beginning of the huge tables that represented the charting of this huge inhospitable land called the Monster. '196,883. That's the smallest-dimensional space in which you can represent this object. The Monster is like some huge great symmetrical snowflake that you can see only when you get to 196,883-dimensional space.'

Escher's chocolate box was a symmetrical object that existed in our three-dimensional world. You could see this object, touch it, play with it. Sitting at the front of the Atlas, it had only 60 different symmetries. Spanning pages and pages at the end of the Atlas was this vast creature that required you to enter 196,883-dimensional space before you 'saw' it. Of course, you could never see this object in a visual sense.

One of my most exciting revelations in the previous years had been how the language of mathematics provides alternative ways of 'seeing' the world. Escher's visual paradoxes reveal how bad we can be at perceiving reality. By changing physical space into the language of mathematics, these paradoxes are easily exposed. Equations allow you to see into the future by making predictions about the flight of a planet or the evolution of the economy. This was a language with far more power for me than the French and Russian I'd battled with at school. But it was the ability of this language to conjure up in the mind's eye things that our physical eyes could never perceive that was for me one of the greatest thrills. Mathematical language opens up a virtual window onto spaces beyond our physical three-dimensional world.

We are actually all used to the idea of turning space into numbers. When we look up the location of a city in an atlas, we find it identified by a grid location. For example, the maths department I visited in Cambridge can be found at latitude 52.2°N and longitude 0.1°E. The same principle is used in mathematics to change geometry into numbers. For example, the four corners of a square can be described by their coordinates: $(0,0)$, $(1,0)$, $(0,1)$ and $(1,1)$. And similarly in three dimensions: one just adds another coordinate. For example, the eight corners of a cube can be described by eight triples: $(0,0,0)$, $(1,0,0)$, $(0,1,0)$, and so on, up to $(1,1,1)$ (Figure 6). The coordinate $(1,0,1)$ locates or encodes a point on the three-dimensional cube reached by travelling one step east and one step vertically upwards.

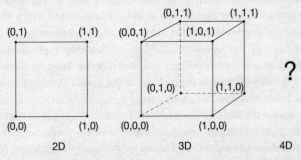

Fig. 6 Changing geometry into numbers: a shape can be described by coordinates.

   The beauty of mathematics is that, now that I have this translation of pictures into a new language of numbers, I can portray the geometry of a cube in four dimensions without having to concern myself at all with trying to visualize it. This four-dimensional figure, known as a hypercube or tesseract, has 16 vertices each described by four coordinates, starting at $(0,0,0,0)$, then $(1,0,0,0)$ and $(0,1,0,0)$, and stretching out to the farthest point at $(1,1,1,1)$. The numbers become a code to describe the shape. Although I can't 'see' the hypercube, the mathematical language allows me to manipulate it and explore its symmetries. The numbers give me, if you like, a sixth sense – the feeling that I really can see in four dimensions.

   Despite my newly acquired ability to 'see' higher-dimensional shapes, Conway and Norton's ability to conjure up a symmetrical snowflake in 196,883-dimensional space was a pretty mind-boggling thought experiment. This was not an object you would see dropping from the sky. To construct such an object you were forced to rely on mathematical language. It existed in a mathematical world where physical objects are replaced by numbers encoding these objects. Just as the hypercube can be described by strings of quadruples made of 0's and 1's, Conway and Norton could pin down the Monster using strings of 196,883 numbers. And, according to Conway, 196,883 wasn't random.

'The amazing thing,' he said to me, 'is that $1 + 196,883 = 196,884$.'

I looked a bit blank. That didn't strike me as something that would get anyone too excited as a great mathematical discovery. 'Ahhh, but 196,884 is the first coefficient in the Fourier expansion of the modular function.' Now, I vaguely knew what this meant. It was something important in number theory. But it was not something which seemed to have anything to do with the symmetry of a huge snowflake. 'That's the point,' Conway countered. 'When someone told me about it it sounded like pure numerology. But then I went down to the library here in the department to find a book about these modular forms. OK – what's the next number on the list?'

I looked at the table. It was 21,296,876, the size of the next important dimension in which you could see this snowflake. 'Well, when I went and looked up the second coefficient of the modular function in this book in the library, it was 21,493,760.' I looked blank again. 'The point is that $21,493,760 = 1 + 196,883 + 21,296,876$. Simon and I found a way to use all the numbers coming from the table for the Monster, to get all the terms in the Fourier expansion of the modular function.'

The point was that this strange thing called the modular function can essentially be described by a sequence of numbers starting 196,884, 21,493,760, 864,299,970, . . . Similarly the contours of this monstrous snowflake were defined by another sequence of numbers: 196,883, 21,296,876, 842,609,326, . . . Conway and Norton had found a bit of mathematical magic which seemed to miraculously turn one set of numbers into the other.

To the non-mathematically sensitive, this might not sound like much, but I knew enough by now to appreciate that this was weird. It was as if an archaeologist excavating a Mayan pyramid in the jungles of Guatemala had revealed strange patterns only ever seen before in the tombs of Egypt: you would have to infer some connection between the two cultures. Conway's excavations had revealed a similar link between two mathematical carvings: the modular function from number theory and this Monstrous symmetry. The two things didn't appear to have anything to do with each other. Yet the secret of which dimensional space this Monstrous creature lived in seemed to be programmed into the modular function.

'That experience was the most exciting of my mathematical life,' said Conway. But what does it mean? 'That's the point: we don't understand it. Why is there a connection?' 'Monstrous moonshine,' Simon Norton chipped in. 'That's what we called it, this strange numerology,' Conway explained, 'Monstrous moonshine'.

It was an intriguing name which immediately caught one's attention. But what sense of the word 'moonshine' were they referring to? The name given to the illegal production of whiskey? Was this connection so strange it was hard to swallow? 'Well, the whole subject is vaguely illicit!' Conway admitted. Or perhaps moonshine was being used to indicate that they were speaking complete nonsense. But this seemed to be more than mad numerology. You might have legitimately dismissed as some strange coincidence the observation that 196,883, the first dimension in which you can see the Monster, and 196,884, the first number in the modular function, were so close. But it had to be more than numerological nonsense that *all* the numbers that Conway and his crew had documented in the Atlas to help navigate the symmetries of the Monster were so directly connected to the numbers coming out of this object in number theory. 'The connections are just too astonishing to be accidental.'

Indeed, what they seemed to be getting at with their use of 'moonshine' was that there appeared to be a kind of mathematical sun whose rays were illuminating the numbers in the Monster and the modular function from number theory. Although we could see the reflected moonlight, no one could see the sun which was the source of the connection between these numbers. The source of this moonshine, Conway said, was one of the greatest mysteries in the subject. I could see the appeal of the problem. The strange interconnected nature of mathematics was one of the aspects of it that I'd begun to find most intriguing. Finding the tunnel between these two subjects, the Monster and the modular function, looked a fascinating project. Like Bottom in *A Midsummer Night's Dream*, who could resist the mathematical weaver's call to 'Find out Moonshine'?

And then, as if I wasn't there, the two of them started bandying round bigger and bigger numbers, coordinates that they'd documented in their Atlas, as they explored more and more of the strange implications of this moonshine. This object was so familiar to them that they had no need to look at the chart open in front of me. They lived the Monster. It was a friend, someone they knew intimately. But this creature was keeping some of its secrets close to its chest, despite the probing questions Conway and Norton were firing at it. I sat there in awe at their ability and command of something so complex that it seemed to lie beyond the capacity of a normal mind. But just as Conway had found clues in the decimal expansion of π to help him remember so many digits, the Monster, despite its size and complexity, had given up enough of the secrets of what made it tick for Conway and Norton to find a way in.

After a while sitting listening to the two of them firing numbers at each other in a mathematical duel, I quietly took my leave. I followed the instructions that Norton had plucked from his plastic bags for the best route back from Cambridge to Oxford.

Midnight, 26 August, the Sinai Desert

At last the temperature has dipped to something bearable. I'm lying out on the sand with the night sky burning above me. I still get a real thrill just looking up into space and wondering what's out there. What shape is it? What does it mean to say that the universe is 'unbounded yet finite'?

I must admit that I am actually a little stoned, thanks to a birthday present from our Bedouin host. The grass grown on the other side of the mountain by the Bedouin is some of the best in the Middle East. Perhaps it's a bit sad of me to be pretending to be as hip as the two other kids we are sharing the beach with. Perhaps it's just me trying to deal with the crisis of hitting 40. At university I'd prudishly passed on joints, convinced that however good they might be for inspiring poetry, they were bad for the mathematical mind – although I've subsequently discovered several mathematicians who've produced their best work under the influence.

The moon has just risen over the mountains of Saudi Arabia. Why does it look so much bigger here than it ever does in London? Is there some strange lensing effect that the atmosphere has here which magnifies the moon? The moon is ageing, in its last quarter. For the

Bedouin, the phases of the moon control the cycles of their year, the crescent new moon marking a new month. According to my hosts, my birthday falls this year in the month of Rajab. Next year my birthday will have crept into another month of the Islamic year. It's the power of mathematics which gets you from one date to the other, although ultimately the authorities across the water in Saudi Arabia have the final say on the Islamic calendar.

The waves are gently lapping over the coral reef. The moonlight is glistening off the surface of the sea. Those photons of light have been on an extraordinary journey. Launched from the sun that set behind me, they've bounced off the moon and hit the surface of the sea before finally landing in my eye. But what actually happens to that photon once it's entered my eye? What's the strange mix of physics and biology that gives me the sensation of seeing the shimmer on the waves?

The moon has been pushing and pulling the sea all day. The tide has turned again, and has now covered the coral shelf where I saw the symmetrical starfish this afternoon. Why are there two high tides a day rather than just one? It's a question that has quite a subtle answer, I realize as I try to work it out, drawing pictures in the sand of moons orbiting the Earth. Science progresses because of the questions we can't answer. Without unsolved problems to work on, mathematics would die. Eventually, I give up on my sketches in the sand. The mysterious moonshine lights my way back to my shack, glowing beneath the stars.

September: The Next Roll of the Dice

Keep a gamester from the dice, and a good student from his book, and it is wonderful.

WILLIAM SHAKESPEARE, *The Merry Wives of Windsor*

1 September, Stoke Newington, London

September for me has always been a month of beginnings. Ahead lies the cycle of the academic year with all its promise of new things to learn and discover. I've just walked my nine-year-old son, Tomer, to school. It's the first day back after the summer holiday. I use the time to drum in the dreaded multiplication tables. Tomer tries to find tricks to work each one out, using a few simple calculations to create a larger body of knowledge he can refer to for the answer.

At school, he is expected to learn his tables by heart, to perfect automatic responses so that he can move on to other things. As we walk down the road together, I try to ignite the mathematical flame in him. I ask him multiplication questions but I don't just run through each table. I ask for 4×4, then 3×5; 5×5 then 4×6; 6×6 followed by 5×7. After a while he spots the pattern: the second answer is always one less than the first. I hope he might get excited as I explain that this will be true whatever numbers I choose. 'OK, we're here. Can you shut up now?' he says, mortified that he might be caught talking maths with his dad.

Walking back home, my brain is already starting to chew over the problem I'm currently working on. I like to work from home. I find an office an oppressive space, constantly reminding me that I'm not having any great ideas. I hate the accusatory way in which the whiteboard stares at me, asking why I haven't covered it in meaningful equations. My preferred canvas is yellow legal pads. Somehow yellow is the right background for mathematics – perhaps it was seeing all those yellow books in Blackwell's as a kid that has made me associate the colour yellow with doing maths. The binding of the legal pad ensures that some semblance of order is maintained against the back-drop of my chaotic thought processes.

I discovered these pads when I was visiting Israel. I now have boxes and boxes of them sitting in the cellar. Since Hebrew is written from right to left the margins are on the right-hand side, but interestingly, irrespective of whether the language is written from right to left, left to right or top to bottom, mathematical equations always begin at the left and flow to the right. At the moment, I've only got the left-hand side of my equation. The right is yet to be filled in.

Most of the time I just sit there doing nothing, getting nowhere. My room at home is a space where I can easily drift in and out of ideas without feeling guilty. It is an extremely messy place. Half the time this gets me down. But it actually is a good reflection of my thought process. I'll start looking for a book buried somewhere deep in the pyramids of paper that sit atop my desk. But during the search I'll often come across something I hadn't been looking for which can take my thought process off in an unexpected direction. By maintaining an untidy room I'm raising the likelihood of making these random connections. Whenever I tidy everything up, the potential for getting random ideas gets filed away too.

I listen to a lot of music when I do mathematics. It helps to make me feel less anxious when I'm not getting anywhere because my mind gets drawn to the music. Sometimes I'll take a complete break and go and tinkle on the piano. I'm an extremely bad pianist. I'm playing through Bach's *Goldberg Variations* at the moment, but at about a tenth the speed they should be played – that's how long it takes me to get my fingers round the next notes. I have this fantasy that music is actually stimulating the same part of the brain that I need for mathe-matics. So perhaps my efforts are acting as a mental workout to keep the neurons fit for my next mathematical assault.

The other stimulant in addition to music that I use (or abuse) at home comes from my espresso machine. The ritual of coffee making forms an important skeleton around which to frame my working day. Paul Erdős, one of the great characters in my subject, once said that a mathematician is a machine for turning coffee into theorems. A few years ago I gave up coffee as one of my new year's resolutions – and I proved nothing of significance during that whole year. So maybe there is some truth to Erdős's quip. Sherlock Holmes used to measure the difficulty of a problem by the number of pipes he needed to smoke to solve it. My measure comes in espresso shots. However, the theorem I'm trying to prove seems likely to require the annual bean output of a minor South American state before it will reveal its secrets.

Conway's Atlas might list all the building blocks of symmetry, but there is still very little understanding of what can be built from these atoms. Part of the research I do is to see what symmetrical objects can be concocted from these indivisible symmetries. It's as though chemists were to take atoms of sodium and chlorine, say, and ask what sort of compounds they could synthesize from these elements.

I've taken one of the simplest of the building blocks: the rotational symmetries of regular two-dimensional shapes such as a triangle or a pentagon. I am ignoring the reflectional symmetry. If necessary, I could play the same trick that Escher did when he destroyed the mirror or reflectional symmetry of the chocolate box that Conway so liked: paint a starfish on the shape with a little anticlockwise twist to its tentacles.

A pentagon has five rotational symmetries. You can turn the shape through $1/5$ of a whole rotation, or $2/5$, $3/5$ or $4/5$, or leave the pentagon where it is. Similarly, the equilateral triangle has three rotational symmetries. In fact, for any regular two-dimensional polygon, the number of rotational symmetries is the same as the number of sides.

So, a 15-sided regular polygon has 15 rotational symmetries. But now something interesting happens. The symmetries of the 15-sided figure are actually built from symmetries of two smaller shapes: the pentagon and the triangle. If I draw a pentagon and a triangle inside the 15-sided figure, I can achieve every rotation of the larger shape by combining rotations of the smaller shapes.

For example, in Figure 7, how can I rotate the 15-sided figure by $1/15$ of a turn, so that A moves to B, by combining rotations of the triangle and the pentagon? If I rotate the pentagon by $1/5$ of a turn, A goes to C. If I repeat this and rotate the pentagon again by $1/5$ of a turn, C goes to D. The last step is to rotate the triangle sitting inside the 15-sided figure anticlockwise by $1/3$ of a turn, to send D to B. The combination of rotating the pentagon twice and then pulling back the other way with a single rotation of the triangle has achieved a turn by $1/15$ of the big shape. It works because $1/15 = 2/5 - 1/3$.

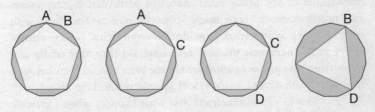

Fig. 7 How to rotate by 1/15 of a turn, using the symmetries of a triangle and pentagon.

You cannot break the pentagon or the triangle into rotations of smaller shapes, as you can with the 15-sided shape. The reason is that 5 and 3 are both prime numbers. Prime numbers are numbers that cannot be written as two smaller numbers multiplied together (with the exception of 1, which is not regarded as a prime). So here are the first and simplest building blocks in the periodic table of symmetry. If you take a regular two-dimensional polygon with a prime number of sides, then the rotational symmetries of this prime-sided shape cannot be built from those of smaller symmetrical objects.

Not only that, but these prime-sided figures are the building blocks for the symmetries of all the other regular two-dimensional polygons. For example, the symmetries of a 105-sided figure come from symmetries of a triangle, a pentagon and a heptagon sitting inside it. This is a geometric way of saying that every number is built by multiplying primes together. This is why the primes are so important, because they are the building blocks of all numbers. When we turn to the mathematics of symmetry, we find that prime numbers are also the building blocks of some of the simplest of the symmetrical shapes.

But although they are the simplest of the building blocks, the assortment of symmetrical objects that can be built from the rotational symmetries of these prime-sided polygons is still an utter mystery. I have become rather obsessed with trying to discover what happens if you take, for example, lots of equilateral triangles. What different mathematical objects are there that have $3 \times 3 \times 3 \times 3 \times 3$ symmetries?

Such an object has symmetries built from piecing together the
symmetries of five triangles. What if I change from triangles to pentagons? And heptagons? There are infinitely many prime numbers and
hence infinitely many prime-sided building blocks. What is the nature
of all the different shapes I can construct whose symmetries are built
from copies of one prime-sided shape? In particular, if p represents
any prime number, how many different objects are there with
$p \times p \times p \times p \times p$ symmetries? How do the symmetrical objects change
as I vary the particular prime I am using? Do they vary wildly as I
move from one prime to another? Or are there connections between
the objects with $41 \times 41 \times 41 \times 41 \times 41$ symmetries and the objects with
$73 \times 73 \times 73 \times 73 \times 73$ symmetries? And what happens when I increase
the number of prime-sided shapes?

Perhaps it's worth flagging up a little warning here. The 'objects'
I'm interested in aren't necessarily physically built out of triangles.
Although I've started with a simple two-dimensional shape, most of
the objects I'm constructing can't be realized in two- or even three-
dimensional space. They are four- or five- or higher-dimensional
objects, and I need the language of mathematics to construct and
manipulate them. What is important is that the total number of symmetries in the object is a power of 3. The symmetries in the object will
thus have been built from rotations of triangles.

Another warning. Even if something really is physically constructed
from triangles, that doesn't mean that its symmetries come only from
rotations of the triangles. For example, as I learnt in the book my
teacher gave me, there is an object built out of 20 triangles called an
icosahedron (Figure 8). This is the shape that Escher used to build

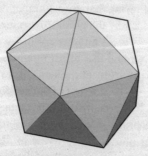

Fig. 8 The icosahedron is built from triangles, but its group of symmetries includes
the rotations of a pentagon.

the chocolate box that Conway described. Since you can make an icosahedron by gluing together triangles, you might legitimately expect that its symmetries are also built from triangles. But there are symmetries in this object that come from a pentagon. For example, each point or vertex has five triangles meeting at it. Rotate the icosahedron by $\frac{1}{5}$ of a turn around a point, and it will look exactly the same.

My research can be compared to a chemist taking atoms of a single element in the periodic table, such as carbon, and asking what molecules you can make from it. The chemists call these different chemicals allotropes of carbon (Figure 9). In fact, symmetry is essential to explaining the different ways that carbon can be pieced together. For example, you can take a carbon atom and arrange four other carbon atoms around it in what is called a tetrahedral arrangement. This makes diamond. The symmetry of the arrangement makes it one of the strongest molecules in nature. Alternatively, you can arrange the atoms in a lattice of hexagons, which makes them look like honeycomb. This makes graphite one of the weakest molecules. Although the two-dimensional hexagonal slabs are quite stable, the honeycomb layers just slip over each other.

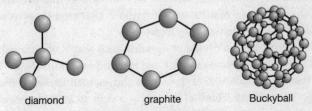

diamond   graphite   Buckyball

Fig. 9 (a) Diamond, (b) graphite and (c) buckminsterfullerene: different ways in which carbon atoms can be assembled.

One of the most exciting stories in chemistry was the discovery in 1985 that 60 carbon atoms could be put together to make a single molecule. The secret to building this molecule, called C_{60}, is to look at the symmetries of a football. A modern football is made up of a patchwork of pentagons and hexagons. The shape has 60 vertices. Harry Kroto, then at Sussex University, and Richard Smalley and Robert Curl at Rice University in Texas realized that it was possible to arrange 60 carbon atoms, one at each vertex, and assemble them to make a new, spherical carbon molecule. They even discovered examples of these molecules in experiments designed to recreate the atmospheres of stars whose outer layers are rich in carbon.

The shape reminded the team who discovered it of geodesic domes built by the architect Buckminster Fuller, so they christened the molecule buckminsterfullerene in his honour. Because of the molecule's resemblance to a football, it is often nicknamed the 'buckyball'. The discovery opened up a whole variety of new ways in which carbon atoms can be put together to create bigger molecules. Again, symmetry was crucial to the understanding of the possible existence of such strange molecules. Once the mathematicians had revealed what was possible, it was only a matter of time before the chemists discovered carbon compounds in nature that exploited these different symmetrical shapes.

My research tries to answer the same sort of question in the world of mathematical symmetry. Instead of carbon, my building blocks are the symmetries of a simple symmetrical prime-sided shape such as the equilateral triangle and the pentagon. What shapes can I create whose symmetries are built from a number of copies of one single prime-sided shape? Again, that warning: bear in mind that what I'm studying are not just three-dimensional forms – they are objects that live only in four, five or more dimensions, but whose symmetries nevertheless reduce to the symmetries of triangles.

Because people can't build or visualize such shapes, the thought of objects in four-dimensional space can be mind-boggling. The art is to find the right language to explore these shapes, even though you can't physically see them. Think of describing a cube to a blind person: by using language, we can convey a sense of the cube by describing how many faces it has, how many edges and vertices.

As I'd discovered during my training at university, turning space into numbers is the most powerful language for describing higher-dimensional objects. Let's take a four-dimensional cube – the hypercube. It has 16 vertices. One of those points is located at a place we can identify in a coordinate system as $(0,0,0,0)$. Four edges emanate from this point. The edges join this corner to four other points which we can identify by the strings of numbers $(1,0,0,0)$, $(0,1,0,0)$, $(0,0,1,0)$ and $(0,0,0,1)$. I can even spin this hypercube through an axis joining the two extreme points $(0,0,0,0)$ and $(1,1,1,1)$. The symmetry has the effect of cycling round the four edges coming out of $(0,0,0,0)$. You'd probably have no problem with following this if the coordinates were those of a three-dimensional cube. To help me 'see' what's happening, I'll often draw a 'shadow' in two dimensions to give some idea of what's happening out there in four dimensions (Figure 10).

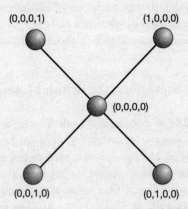

(0,0,0,1)   (1,0,0,0)

(0,0,0,0)

(0,0,1,0)   (0,1,0,0)

Fig. 10 Projecting the corner of a four-dimensional cube into two dimensions.

The language of numbers gives me a way to play with the geometry of an object I'll never be able to build in reality. It might be a little harder because I have no conventional, three-dimensional image of it, but that doesn't make the task impossible. For example, I can 'see' that if I repeat the spinning of the hypercube I described above, after four rotations the hypercube will have returned to its original position. This rotation actually looks like the rotation of a square. So I know that the symmetries of this hypercube are not built from the symmetries of triangles – it's not one of the objects I'm after in today's investigations.

When I explore these shapes, I often feel as though my office is the portal to a magical land. My desk is like C. S. Lewis's wardrobe whose doors lead into another world beyond the coats hanging inside. Sometimes I can spend the whole day trying to get through the wardrobe, but I just can't find a way past the wooden panel at the back. But when the magic works and I find a way in, instead of a Narnia populated by fauns with umbrellas and talking lions, I enter a world containing spinning hypercubes and Monsters lit by moonshine. Just as the children in Lewis's tale found it difficult to get back from Narnia into wartime London, I sometimes get stuck in this mathematical world, dislocated from what is going on around me.

It's 3.45 before I know it. Time to find a way back through the wardrobe and pick up Tomer from school. My head is fizzing from the day spent in this strange act of mathematical meditation, so I'm quite happy to head for the park with a ball made from pentagons and hexagons and wind down with a kick-about with Tomer.

10 September, the British Museum

It's the weekend. I've done a deal with Tomer: a morning in the British Museum looking for symmetry, followed by an afternoon at the skateboard park. My grandparents used to live round the corner from the museum. I always enjoyed staying the night in their flat and waking up in the morning to the rumble of the London streets: police cars, buses and taxis all sounded so exotic to a boy used to the sedate traffic of a town in the Thames Valley. I spent many Saturday mornings as a child wandering round the museum's Greek and Roman galleries, hiring an audio guide and doing the tour of the Elgin Marbles.

I've been spending the last few days looking for 'objects' whose symmetries are built from the symmetries of triangles or pentagons. It mirrors a search which began almost as soon as humans started fashioning their environment for their own purposes. Tools for hunting and pots for cooking exploited the variety of different geometric shapes that could be built out of clay, stone or bone. Some of the first investigations into the symmetry of three-dimensional space were driven by our obsession with games. Tomer and I are off to see if we can find any ancient games that might help me to trace the history of the discovery of different shapes.

In Britain, early humans had fashioned a range of quite sophisticated symmetrical shapes. Five thousand years ago, Neolithic people were setting Stonehenge and other great stone monuments into the landscape of Britain. The placement of the stones shows a fascination with symmetry. The stone circles create shapes on the ground with sometimes as many as a hundred sides. Some of the stones in these circles are very widely spaced – at Avebury in Wiltshire, for example, the outer circle of stones runs for over a kilometre. To build such circles would have required sophisticated mathematical skills, or at least a heightened sensitivity to the creation of something with symmetry.

Early primitive art on pots and walls also reveals this increasing sensitivity to symmetry. On the walls of tombs in Ireland dating from the same period as Stonehenge, spirals are often carved into the stone. Two groups of three spirals greet you as you enter one of the most famous tombs found in Europe, the Newgrange tumulus in County Meath. The three spirals are arranged in a triangle (Figure 11). The spirals in each group wind a different way – one group is a mirror image of the other. On the walls of such tombs, symmetrical symbols abound: concentric circles or squares, a row of diamonds, stars with symmetrically arranged points. Pictures of circles with radial lines are clearly images of the sun. One of the most impressive carvings is the so called Stone of Seven Suns, at Dowth in Ireland. Looking like a set of wheels with an outer and inner circle joined by radial lines, these seven suns are perhaps an illustration of the sophisticated grasp of astronomy that Neolithic man already had 5,000 years ago.

Fig. 11 One of the groups of spirals at Newgrange tumulus in County Meath.

Just as notations for numbers developed to record calendar dates, the huge ·array of symmetrical symbols used in these tombs seem to be different ways of keeping track of time. It is perhaps significant that these people chose symbols of symmetry to represent the natural cycles of the seasons, and the movements of the sun, the moon and the stars. The visual language they were creating mirrors the temporal patterns they had recognized. The kite divided into four quarters (Figure 12), for example, has been suggested by archaeologists to have symbolized the passing seasons.

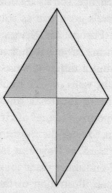

Fig. 12 The kite shape was used as a symbol to represent the year divided into four seasons.

Around this period, along with these stone circles set into the landscape and symmetrical shapes etched on the walls, Neolithic people started to carve a range of interesting three-dimensional shapes packed with symmetry (see Figure 13). Hundreds of balls carved from basalt or sandstone have been discovered in north-east Scotland which date back to 2500 BC. Etched into the sides of the balls are geometric patterns. The sculptors have played around with different symmetrical arrangements of protruding knobs, in a similar way to the patches on a modern football. Over half of these balls have six round patches carefully carved into the stone. Although the ball is round, the sculptor has exploited the symmetries of the cube in arranging the six patches.

The sculptors also found that they could carve four circles into a ball to create a pleasingly symmetrical arrangement, the same layout that nature uses to make diamond out of four carbon atoms. One particularly fine example, found at Towie, is housed in the Museum of Scotland in Edinburgh. A few years ago, during a conference I was

Fig. 13 Neolithic stone balls demonstrate a very early fascination with symmetry.

attending in Edinburgh, I skived off lectures one afternoon to search out this ball in the museum. The ball was larger than I expected, about the size of a fist. The four circles are themselves intricately decorated with a complex pattern of spirals and concentric circles (Figure 14). While my fellow mathematicians presented their twenty-first-century explorations of symmetry, I marvelled at this beautiful stone whose symmetry began the journey. The sculptors also carved a variety of other arrangements, including as many as 12 or 14 knobs, and even one with 160. They could not have failed to notice that 12 knobs were easier to arrange than 10 or 14 (but explaining why this was so would take another three millennia). The artists may also have decorated the patches with different colours, which would have highlighted some of the different symmetries of each carving.

It is not clear what role these stones played in the culture of Neolithic

Fig. 14 A carved symmetrical ball found at Towie in Scotland.

Scotland. Some have suggested that they were used as symbols of authority by clan leaders. They have never been found in tombs, so they may have been more significant to the tribe than to any particular member of the group. Some of these geometric patterns on the carved balls have also been found engraved on objects such as mace heads. Perhaps symmetry was being appropriated by the ruling class as a symbol of authority.

The advent of gambling and dice games in many cultures across the world in the first millennium BC pushed different civilizations to explore what shapes make the best dice. Symmetry is essential if you are going to make an object which is as likely to land equally on any of its faces when thrown to the ground. The first dice were not six-sided but four-sided, and made from the bones of animals. The ankle bones of a sheep, known as knucklebones, are shaped in such a way that they naturally fall in one of four ways. These early examples of dice have been found at many prehistoric sites. But it became clear that these bones when used as dice are likely to be biased towards one side, and ancient cultures were soon looking for ways to sculpt bones to make for a fairer game.

Tomer and I couldn't find any of these Neolithic dice or knucklebones in the British Museum. But we did have more luck in our search for symmetry in the museum's Mesopotamian gallery, where we discovered an intriguing board game with a set of pyramid-shaped dice. The board itself is inlaid with symmetrical patterns made from shells, lapis lazuli and limestone. There are two regions, one with twelve squares and the other with six, joined by a bridge of two squares. Each square bears its own symmetrical symbol which denotes the significance of its location. There are eight-leafed rosettes of red and blue, diamonds and squares.

The game dates from 2500 BC and was found during an excavation of the ancient city of Ur, in southern Iraq. A Babylonian cuneiform tablet dating from 177 BC gives a partial description of the rules of the game. It seems that the squares with the most symmetrical image of the rosette were regarded as the lucky spaces that the players would aim for.

It was the four dice, however, that most intrigued us (Figure 15). Each die was a tetrahedral pyramid with four faces of equilateral triangles. But unlike the dice Tomer and I use when we play Monopoly at home, on which each face has a value, these dice worked slightly

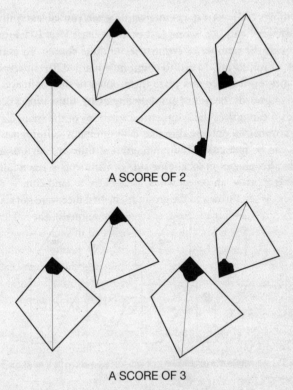

A SCORE OF 2

A SCORE OF 3

Fig. 15 The score with the tetrahedral dice in the game from Ur corresponds to the number of dots that point upwards after a throw.

differently. I wasn't sure, peering through the cabinet, but it looked as though one vertex or point on each die was marked with a dot. When you threw the stones, the score of the dice would correspond to the number of dots that appeared on the top of each die. So you could get any score between 0 and 4.

What were the chances of getting each of the possible scores? It's the sort of question I can't help asking as soon as I see such a puzzle. Tomer could see what was coming too. 'What is the most likely throw?' Tomer's eyes rose heavenward. Each die has a 1 in 4 chance of landing on the face that leaves the dot pointing up. So there's only going to be a 1 in $4^4 = 256$ chance of rolling a 4. On the other hand there is a $3^4 = 81$ in 256 chance of scoring zero and missing your turn.

But things really start to get interesting when you look at how many different ways you can score 1, 2 or 3, because that is inextricably linked with the number of symmetries of these shapes. To assess the chances of rolling a 1, I need to count how many different ways there are to pick up the dice and place them down so that only one dot is pointing upwards. Because I am only interested in the vertex pointing upwards, I can ignore the symmetrical rotations of the triangular face. So the symmetries of these dice that determine the probabilities of the throws can be reduced to the symmetries of four squares sitting on a rod like a combination lock. One side of each square has a dot on it (Figure 16).

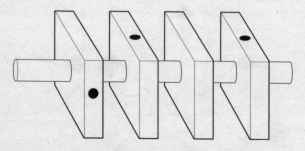

Fig. 16 The symmetries of a combination lock with four squares can be used to analyse the dice in the game from Ur.

The symmetries of the combination lock are a subset of the symmetries of the four tetrahedra. Since the number of different combinations is $256 = 2^8$, a power of the prime 2, the group of symmetries of this combination lock is actually one of the objects that I have been trying to understand sitting at home in my office in Stoke Newington. To calculate the chances of getting a 1 from these four dice, I need to calculate how many symmetries of the combination lock leave one dot facing outwards.

I start by choosing the dot showing on the first wheel. That's like saying the first die landed spot-up. How many different ways are there for the other three dice to land with no spot showing? Each of the other wheels can be turned in three different positions to show a blank side. So that is a total of $3 \times 3 \times 3$ configurations. But I could also have chosen the dot to be showing on the second or third or fourth wheel.

Again, the remaining wheels can be arranged in $3 \times 3 \times 3$ ways to show blanks. So that makes a total of $4 \times (3 \times 3 \times 3) = 108$ out of 256 symmetries that give me 1. If you calculate the other possibilities you get a 54 in 256 chance of getting 2 and a 12 in 256 chance of getting 3. So the most likely throw with these four dice is 1.

Tomer is looking somewhat glazed. But I rather like this conceptual leap, where the problem suddenly becomes tractable by visualizing it in a completely different way. Instead of throwing tetrahedra, I've changed the problem into spinning combination locks. 'It sounds like that problem we did in Guatemala when you turned getting to the supermarket into a problem about necklaces,' says Tomer. Two years ago we spent seven months living in Guatemala. We were living in Antigua, one of the first towns to be built as a grid of *avenidas* and *calles*: seven parallel streets crisscrossed by seven parallel avenues (Figure 17). Our house was in the top right-hand corner of the grid. The supermarket was at the bottom left. We spent some time trying

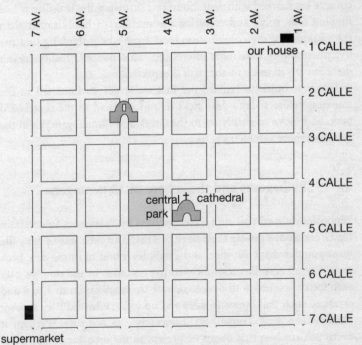

Fig. 17 How many ways are there to get from our house to the supermarket?

to work out how many different ways there were to get from home to the supermarket.

Tomer found that when the problem involved just three avenues and three streets, there were six different routes. But drawing paths through the city was not going to solve the bigger problem. By changing the problem into counting how many necklaces you can make out of six red beads and six yellow beads, I managed to crack it. Each necklace represented a path through the town – a red bead meant go west; a yellow bead, head south. But counting necklaces turned out to be much easier. In the end, by using the formula I came up with we calculated that there were 924 different routes through the town – enough to keep us going for a couple of years before we had to repeat a route.

As we're about to leave the cabinet containing the Game from Ur, Tomer suddenly points at the dice. 'Dad, look! There are two dots on the dice, not one.' Sure enough, when I look closer, two of the four corners are marked with dots. So in fact throwing the real dice is like flipping four coins and counting the number of heads rather than the workings of my combination lock. I can't help thinking that my hypothetical dice were more interesting, but sometimes reality doesn't quite live up to one's mathematical expectations.

Tomer is quick to remind me that we've got an hour left in the museum before today's deal kicks in and we head to the skateboard park. So I move us swiftly on to the Greek and Roman galleries in the search of more symmetrical dice.

Pythagoras and the sphere of 12 pentagons

Sophocles claimed that dice were invented by Palamedes to entertain the Greek troops during their siege of Troy. The cube was by now the most popular shape for dice, and examples found in Rome date back to 900 BC. These Etruscan dice are very similar to the one we play with today, marked with dots to denote the numbers from 1 to 6 and arranged such that opposite sides add up to 7. 'Why did they arrange the dots like this?' Tomer asks innocently. I'm not sure. 'Perhaps it evens out any bias that might be present in the dice due to imperfections in making a perfect cube,' I suggest tentatively.

Roman soldiers were so obsessed with dice games that they would carry heavy dice boards with them on their backs along with all their military equipment. In around 500 BC a new shape appeared on the scene. This new die had 12 faces as opposed to the six of the cube. The faces were in the shape of a pentagon rather than the squares of the cube or the triangles of the pyramid dice in the game from Ur. The Romans found that these 12 pentagons could be carved out of a ball of stone in such a way that no face was favoured over any other. The symmetry made it an ideal candidate for a new die. Examples of these dice have been unearthed near Bologna with Etruscan-Roman numbers carved onto the 12 faces of the shape.

This new die made up of 12 pentagonal faces is quite a sophisticated shape. It is not at all obvious without seeing the shape assembled that 12 pentagons can be pieced together so symmetrically. The Romans may have discovered these 12-sided symmetrical shapes because they were familiar with fool's gold – a compound otherwise known as pyrite which often arranged itself into eye-catching crystals. It is often found alongside copper, and miners would have been used to seeing it in both its cubic form and in large lumps made up of crystals with pentagonal faces. The crystals are not completely symmetrical and wouldn't be suitable as dice. But they might have provided the inspiration for the Roman sculptors who discovered that you could actually level off the sides of a pyrite crystal to make each side a perfect pentagon.

In 500 BC, mathematics had yet to crystallize into an independent discipline. Both the Neolithic stones carved in the third millennium BC and the Roman dice were experiments in symmetry rather than the products of any well constructed theory. These cultures were simply picking up and playing with the range of interesting shapes they saw around them. It was the arrival of a Greek mathematician in southern Italy at this time that marked the beginnings of a more analytical approach to the world around us.

Pythagoras was born on the island of Samos around 570 BC. As a young man he was encouraged by his elders to spend time studying in Egypt. While he was there the country was invaded by the Persians, and Pythagoras was taken prisoner and shipped off to Babylon. His travels were very influential in shaping his mathematical view of the world. The Egyptians instilled in him a strong sense of geometry, while in Babylon he picked up their sophisticated arithmetic skills. From both cultures he acquired a deep sense of mysticism which infected much of his later work.

Eventually Pythagoras returned to Samos and tried to found a school around him which would draw on the symbolic and geometric approach to the world that he had come across on his travels. He believed that the reality surrounding him was bound together by mathematical ideas. He also believed in the important spiritual significance of certain symbols such as the pentagon and the triangle. His ideas were not received well by his fellow Samians, so he moved instead to Croton, on the southern tip of Italy.

It is here that he came into contact with the symmetrical shapes being utilized by the dice-crazed Romans. He developed a fascination with the die with 12 pentagonal faces. Already obsessed with the mystical significance bound up in the pentagram, he must have felt a thrill on discovering that 12 pentagons can be perfectly arranged to make an object full of symmetry. None of the 12 pentagons is favoured over any of the others.

The motion of the moon across the sky naturally divided the year into 12 months. So the discovery of this object with 12 faces must have appealed to Pythagoras's sense of mathematical mysticism. Indeed, on some artefacts of this shape recovered by archaeologists are carved the 12 signs of the zodiac. The 'sphere of 12 pentagons', as this shape first became known, assumed a spiritual significance for the Pythagorean sect that grew around this Greek mystic. Medieval and Renaissance texts mention the use of 12-sided dice for divination in ancient times.

Along with the sphere of 12 pentagons, the Pythagoreans recognized two other objects as important cousins. The cube, composed of six squares, was clearly an object with symmetry that deserved a place alongside the sphere of 12 pentagons. So too did the four-faced triangular pyramid, the shape of the dice in the game from Ur. The Pythagoreans called this pyramid shape a tetrahedron, *tetra* meaning 'four'. Although Pythagoras cannot be credited with discovering any new symmetrical objects, it was his sensitivity to the symmetry of these three polyhedra that led him to group them together as examples of a common species. In his eyes they were manifestations of a deeper mathematical idea. This first inkling of a move to abstraction would set mathematics apart as a new discipline distinct from philosophy, religion or science.

Pythagoras could see that the abstract bond between these three shapes did not extend to another shape he was familiar with from his time in Egypt. The pyramids of Giza were built during the same period that the Neolithic people of Britain were erecting their great stone circles. Sitting in the desert, the pyramids might have looked as symmetrical as the Pythagorean tetrahedron (Figure 18). But as soon as one begins to analyse the shape, it becomes clear that it lacks the same level of symmetry as the other shapes. The Egyptian pyramids are built from four triangles and a square. The resulting mathematical pyramid is a shape whose symmetry is restricted to that of the square at its base.

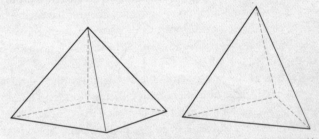

Fig. 18 A square-based pyramid has less symmetry than a triangular-based pyramid.

If you play with the two pyramids, the square-based Egyptian pyramid and the triangular-based tetrahedral pyramid, then you soon get to see why the tetrahedron has more symmetry. I can roll it onto any of its other three faces and put it back down in the desert and it won't look any different. The square-based pyramid has to be replaced on its square if it is to look the same. There are four different ways to set it back down, but beyond that the rotational symmetry of the shape is rather limited. The rotational symmetry in the tetrahedron, on the other hand, gives me 12 different ways I can replace it on its triangular plot. Similarly, the tetrahedron has more reflectional symmetry than the square-based pyramid.

While the Pythagoreans recognized a common theoretical bond between the cube, the tetrahedron and the sphere of 12 pentagons, it is intriguing that they failed to identify another important symmetrical polyhedron that deserves its place alongside the three Pythagorean shapes. The missing object can actually be built from the square-based pyramids of Egypt, that are close to perfect equilateral triangles.

Although a single Egyptian pyramid is not so exciting symmetrically, the shape comes into its own when you take two of them and glue them together along the square bases. If the triangular faces of the pyramids are all equilateral triangles, the object you get is rather remarkable (Figure 19). It can be viewed from many different angles and still appears the same. As you turn it, it becomes impossible to tell along which plane the two original pyramids were fused together. It is built from eight triangles. At each of its points four triangles meet, whereas in the tetrahedron three triangles meet at each point. Here was a new shape, built from the same triangles as the tetrahedron but put together to create an eight-sided figure called an octahedron.

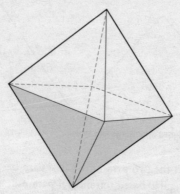

Fig. 19 The octahedron, built from eight equilateral triangles.

If you don't use equilateral triangles to construct an octahedron, it will have less symmetry as it will be slightly squatter or elongated in one direction, according to the choice of triangle. For example, the square-based pyramids that Tomer and I visited when we were in Guatemala were built for height. The builders were, after all, aiming to reach above the canopy of the jungle. So two Guatemalan pyramids fused together would clearly lose the symmetry.

This is one of the features of the pyramids in Egypt that make them so remarkable. They are nearly always designed with sides that are close to perfect equilateral triangles and thus give the impression of a regular octahedron with one half buried in the ground. So although outwardly the pyramids at Giza don't look as symmetrical as a tetrahedron, their majesty owes much to the symmetry suggested by the shape. Indeed, in the searing heat of the desert, a mirage might well have created the illusion of an octahedron floating in the air. The Egyptian geometers must have been aware of the mathematical beauty and significance of their design.

It is intriguing that the Pythagoreans failed to elevate the regular octahedron to the status that they accorded the triangular-based pyramid, the cube and the sphere of 12 pentagons. They must have known about this polyhedron. Many crystals have this shape: for example diamond when it is mined from the ground is octahedral, as is a red crystal called spinel, often mistaken for ruby. Although the Pythagoreans had made the first move towards abstraction by recognizing their three shapes as examples of a common species, they probably hadn't yet made the conceptual jump to understanding the common feature that underpinned them.

The Pythagoreans believed that their insights into the mystical world of mathematics were so precious that members of the sect were sworn to secrecy. Indeed, the Syrian philosopher Iamblichus, writing in AD 300, claimed that the Pythagorean Hippasus of Metapontum was drowned at sea for revealing secrets about the sphere of 12 pentagons. Other commentators, however, put Hippasus's death down to the fact that he leaked the discovery that the square root of 2 cannot be written as a fraction. This mysticism and secrecy could explain why another century passed before it was realized that two more shapes full of symmetry, the octahedron and the 20-faced icosahedron, deserved their place alongside the cube, tetrahedron and sphere of 12 pentagons.

There was also another reason for the Pythagoreans' failure to capitalize on their early mathematical successes. When the sect began to mix politics with mathematics and mysticism, they ran into trouble. In 460 BC the Pythagorean brotherhood was violently suppressed: its meeting houses were sacked and burned, and many of its members were slain.

Plato – from reality to abstraction

It was the Greek philosopher Plato who picked up the mantle of the Pythagoreans. A century after Pythagoras's death, Plato states in his great work the *Republic* that the study of the Pythagorean shapes had become sorely neglected. The *Republic* takes the form of dialogues between Socrates, Plato and other characters. One part sets out the subjects that are essential knowledge for those who are to lead the state. Arithmetic and plane geometry are considered to be vital skills, not only for their usefulness in war but also because the eternal nature of the concepts 'will draw the mind to the truth and direct the philosopher's thought upwards'.

Astronomy is about to be put next in line, when Socrates interjects: 'More haste, less speed. In my hurry I overlooked solid geometry because it is so absurdly undeveloped. The neglect of solid geometry would be made good under state encouragement.' The word 'geometry', which literally means 'measuring the earth', had been reserved for the mathematics of navigation and mapping, but Socrates saw geometry as going beyond simple measurement. During these discussions there is an interesting tension which emerges between the practical reasons for studying geometry and arithmetic and the pure pursuit of truth – a tension which still runs through mathematics today.

Plato took up the gauntlet thrown down by Socrates to begin some systematic study of solid three-dimensional shapes whose symmetries mark them out as eternal objects that draw the mind to deeper truths, as Socrates had hoped. But the messiness of human affairs forms the backdrop to these eternal discoveries, and Plato's disillusionment with politics, especially after the execution of his mentor Socrates, impelled him to leave Greece for a period.

As happened to Pythagoras before him, Plato went to Egypt and the influence of the ideas he found there helped to shape his strongly geometric view of the world. In 387 BC he returned to Athens to set up an institution devoted to research and the teaching of science and philosophy. He hoped to put into practice his dialogues with Socrates on what constituted suitable training for the next generation of political leaders in Greece.

Plato's institution was founded on a piece of land that had once belonged to a mythical hero called Akedemos, and in his honour it became known as the Academy. It was during discussions at the Academy that Plato's friend Theaetetus began to understand the principle that underpins the solids that the Pythagoreans had held dear: the cube, the pyramid and the sphere of 12 pentagons. Plato described his friend as having a snub nose and protruding eyes but a mind of beauty. And it was for capturing the abstract mathematics of symmetry that he would ultimately be remembered.

Theaetetus could see that if you wanted a three-dimensional shape with lots of symmetry, then it was important to build it from two-dimensional polygons that were symmetrical. And if all the faces were the same shape, that would potentially increase the symmetry of the resulting solid. But he recognized too that there were limits to the types of polygon that can be used. As the bee discovered, hexagons can be put together to make only a flat surface. A die with hexagonal faces is impossible. Shapes with more sides than the hexagon don't fit

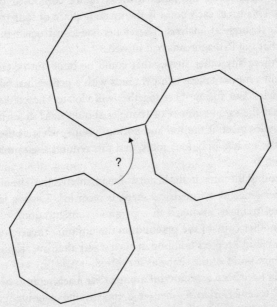

Fig. 20 It is impossible to piece together polygons with more than six sides to completely cover a surface.

together at all. Put two together, and there just isn't room to squeeze in a third face of the same shape (Figure 20). So the faces would have to be built from shapes with fewer than six sides.

So what shapes can you build out of triangles, or squares or pentagons? To achieve as much symmetry as possible, Theaetetus reasoned, the flat faces should always meet each other in the same configuration. No point or edge should look any different from any other, or it would break the perfection of the construction.

The cube certainly met this criterion: six square faces with three squares meeting at each point. The sphere of 12 pentagons was made of 12 regular pentagons, again each meeting three at a time at the points on the shape. As for equilateral triangles, the Pythagorean triangular-based pyramid was built from four triangles meeting three at a time at each point. But with his abstract criterion for selecting shapes, Theaetetus now recognized that equilateral triangles could be configured in an alternative way to create another symmetrical object that was on a par with the Pythagorean polyhedra: two square-based pyramids fused along the bases made a figure composed of eight triangles. This time, each point is the meeting place of four triangles. Using his theoretical analysis, Theaetetus had constructed the octahedron that the Pythagoreans had missed.

Were there any other shapes that could be built out of triangles? One might propose taking two pyramids with a pentagonal base with five triangles. But fusing these together will violate Theaetetus's condition that the same number of triangles should meet at each point. Five triangles meet at the top and bottom points, while at the points around the middle only four meet, and this reduces the symmetry of the object.

It is around this time in the history of mathematics that the discovery of another amazing symmetrical shape is recorded. Starting with five equilateral triangles arranged in a pyramid configuration, instead of gluing another copy of the pyramid on the bottom, Theaetetus found that you could keep on building in such a way that five triangles met at each new point of the shape as it evolved.

It must have been a wonderful moment for Theaetetus as he started to piece together triangles five at a time round a point, gradually building and seeing the shape evolve into a perfect regular polyhedron with 20 triangular sides (Figure 21). It's not obvious at first sight that this object should exist – it's not a shape that anyone had seen in the natural world. You can't piece together 19 or 18 triangles to create a shape where all the triangles meet in such a symmetrical manner. An act of mathematical creation had brought this 20-sided figure into existence. Eventually it would be realized that this shape does occur in nature, but only once microscopes allowed us a closer view of the world.

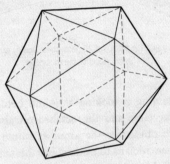

Fig. 21 Twenty equilateral triangles can be pieced together to make an icosahedron.

The Ancient Greeks called the shape an icosahedron, meaning '20 faces'. Another way in which the Greeks might have discovered this new shape is via a close bond the icosahedron has with the sphere of 12 pentagons. This 12-sided polyhedron has 20 vertices. If an equilateral triangle is positioned at each of these vertices and all 20 triangles are suitably oriented, they match up perfectly to produce the icosahedron. If you count the vertices of this new shape built from triangles, you find there are 12. Put a pentagon on each of these vertices and you get the sphere of 12 pentagons back (Figure 22). This close bond between the two shapes mathematicians call duality. You can play the same trick with the square and the octahedron. But if you try it with the tetrahedron, all you get is another tetrahedron.

This duality is important because it actually explains why the dual shapes, although physically very different, have the same symmetries. The symmetries of the sphere of 12 pentagons are the magic trick moves, the things that can be done to the shape which leave it looking like it did before the move. By constructing an icosahedron around the sphere of 12 pentagons, with a triangle at every vertex, one finds that the moves that preserve the sphere of 12 pentagons also spin the

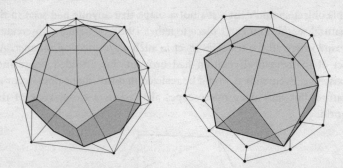

Fig. 22 The icosahedron and the sphere of 12 pentagons. Duality provides a passage from one symmetrical shape to another.

icosahedral case, leaving it looking as it did before the move. The identification of the cube and the octahedron also reveals that the symmetries of each object are the same. It would take another two millennia for mathematicians to understand this subtle idea of underlying symmetries. The Greeks were only just starting out on the abstract journey to explain the common symmetrical bond connecting all these shapes.

If Theaetetus hadn't constructed the icosahedron, it wouldn't have been too long before someone else made the discovery. Perhaps he wasn't even the first to build it. There is something universal and timeless about this shape. It's easy to forget that someone first built it, so in an almost Oedipal act, the creator's name has been forgotten and the shape goes by its Greek name: the icosahedron, the 20-sided shape. Quite often such discoveries are made simultaneously by several mathematicians independently. Ancient Chinese incense burners built in the first millennium AD have these perfect shapes. The craftsmen found the shapes independently and without any contact with the Greeks.

Mathematical proof

Theaetetus now had five dice in his collection, but were there any more interesting dice out there for the mathematical mind to construct? In one of the first examples of mathematical proof, Theaetetus explained why we will never find a sixth way to put together regular faces to build a new sort of die. The limitations of geometry and symmetry can be exploited to show that in three dimensions these five shapes are the only possible dice that can be built. It is this new power to prove facts about the world around us with 100 per cent certainty that during this period in history marks out mathematics as something genuinely different from the other sciences. This is no longer simple observation and butterfly collecting. The mathematician can look into the future and say categorically that these five shapes are all we will ever be able to build from copies of a single regular symmetrical face.

The first recorded account of the five shapes is Plato's description in his text *Timeus* in which he outlines his creation myth. For him, the five symmetrical shapes were so fundamental that they form the very building blocks of matter itself. The triangular-based pyramid or tetrahedron, the spikiest and simplest of all the shapes, Plato believed represented the element fire. The icosahedron is the roundest of all the shapes made up out of its 20 triangles. It represented the element water in Plato's classification, being the smoothest of all the figures. The other figure made from triangles is the eight-sided octahedron. As a shape intermediary between the first two, Plato believed it represented air. The cube with its six square faces represented the element earth, being one of the more stable of the shapes.

This left the sphere of 12 pentagons unaccounted for. Plato renamed it the dodecahedron to indicate that it had twelve (*dodeca* in Greek) faces. Plato believed that this figure 'God used for arranging the constellation of the whole universe'. Plato's God is definitely a mathematician at heart, and this vision has been instrumental in establishing in Western thought the connection between mathematics and theories of the cosmos. Plato's account of the five regular polyhedra gives us the collective name they go by today: the Platonic solids.

It is the Greek language that assigned a name to the common trait that bound Plato's five objects together: *symmetros*. In the first century AD the Roman author Pliny the Elder bemoaned Latin's lack of a word for symmetry. *Symmetros* combines the Greek words *syn*, meaning 'same', and *metros*, meaning 'measure'. Together they describe something 'with equal measure'. Symmetry for the Greeks was reserved for describing an object in which some of the internal physical dimensions were the same across the shape. In symmetrical solids the edges were all the same length, the faces all had the same area, and the angles between adjacent faces were all equal. Symmetry is about measurement and geometry. It would take some time for symmetry to become recognized as a mathematical property that goes beyond simple measurement, although the Greek philosophers were beginning to explore the idea of symmetry as a powerful image beyond physical shapes.

In his *Symposium*, Plato tells us that symmetry not only holds the secret to the structure of matter, but also explains the origin of love. He presents a debate between some of the great thinkers of Ancient Greece on the nature of love. Having planned a night of drinking but overindulged themselves the previous evening, they decide to delay their party and instead hold a competition to see who can come up with the best explanation of the origin of love. The fourth to speak is Aristophanes, who offers the theory that love comes from our craving for symmetry.

According to Aristophanes, humans were once four-legged, spherical beasts with two faces, one on each side of their head. But Zeus, angered at the arrogance of the human animal, came up with a plan to humble their pride: 'Men shall continue to exist, but I will cut them in two and then they will be diminished in strength and increased in numbers; this will have the advantage of making them more profitable to us.' And he slices all humans in half. And that, according to Aristophanes, is the origin of love – our craving to be united once again as a complete being, a perfectly symmetrical sphere.

Intriguingly, Darwin's theory of evolution supports Aristophanes' view that symmetry is the dominant force in our selection of sexual partners. Even Plato's view of the cosmos, based on the symmetrical solids, shares some ideas with modern scientific models. Although Plato's chemistry, with its elements of earth, water, air and fire, is wrong, the four shapes that Plato associated with them do permeate the microscopic world. But that world would not be revealed until humans had developed tools to see things beyond the shapes of the various artefacts on show in the cabinets of the British Museum. Even Plato's association of the dodecahedron with the configuration of the universe now finds an echo in one of the current theories on the overall shape of the universe.

Tomer and I have come to the end of our Saturday morning visit.

As we pass through the main atrium of the British Museum on our way out, we are still assaulted by symmetry. The new roof there is a lattice of triangles put together like a huge, many-sided die.

I'm eager to return to the mathematical tools I've been developing to see what symmetrical shapes I can cook up beyond the three-dimensional world that the Romans and Greeks were exploring. But Tomer reminds me of our deal. So first it's a trip to the skateboard park, to watch him grind and ollie.

October: The Palace of Symmetry

And I would have the Composition of the Line of the Pavement
full of Musical and geometrical Proportions; to the Intent that
which-soever Way we turn our Eyes, we may be sure to find
Employment for our Minds.

LEON BATTISTA ALBERTI, *The Ten Books of Architecture*, 1755

17 October, en route to Granada

At a conference in Edinburgh a few years ago, John Conway told me
that he sometimes spent hours staring at brick walls. Was this a form
of meditation to free the mind from the pressures of daily life to escape
into the abstract mathematical realm, I asked? Not at all, Conway
replied. As we strode around the university campus he pointed from
one arrangement of bricks to the next, explaining how each illustrated
a different sort of symmetry. You have to look at quite a few walls to
find enough different patterns, but then some of the secrets of symmetry will start to reveal themselves.

If you want to delve further into the variety but also the limits of
symmetry, the walls, floors and ceilings of the medieval palaces of the
Moors are where you should look. The Greeks and Romans had started
to explore the symmetry of shapes that make good dice, and discovered
that there were only five perfectly symmetrical polyhedra you could
build. But the Arabic artists who were decorating the palaces for the
caliphs and sultans of the Muslim world started to push the concept
of symmetry beyond the Greek idea of equal measurements. On the
walls of the medieval citadels they began to play a new sort of sym-

metrical game, competing to come up with ever more sophisticated patterns that would repeat themselves in interesting ways. Starting with simple square tiles and the hexagonal lattice of the beehive, they found a plethora of curious designs.

We can look at the array of different shapes that the artists dreamt up and wonder whether they thought there was no end to the catalogue of symmetries. But just as the Ancient Greeks discovered, there are in fact limits to the symmetrical games played by the Moors. In contrast to the five tales of the Greek Platonic solids, the story of symmetry hidden in these palaces is a saga written in 17 chapters. Each exotic tiling is in fact an example of one of 17 different underlying symmetries. While the Greeks had developed the analytical tools to prove that there wasn't a sixth regular polyhedron missing from their list, the abstract analysis to explain this new story was well beyond the Moors. It would take the sophisticated mathematics of the nineteenth century to come up with a complete understanding of the symmetries of these 17 different designs.

Just as three oranges and three apples are both different manifestations of the abstract idea of the number 3, mathematicians would eventually reveal how two apparently different walls could be expressions of the same underlying group of symmetries. Although the Moors could not prove the impossibility of an 18th symmetry, they did at least manage to produce examples of all 17 possible symmetries.

One palace in particular, built around 1300, has always been a Mecca for those addicted to this part of the mathematical story of symmetry: the Alhambra in Granada. Perched in the foothills of the Sierra Nevada mountains in southern Spain, the town of Granada seems almost to grow out of the fertile plains of Andalusia. Surrounded by luxuriant woods, the Moorish palace sits on top of a hill overlooking the town, like 'a pearl set in a bed of emeralds', as one poet described it.

It has become something of a pilgrimage for mathematicians to come to the Alhambra and, as if taking part in a treasure hunt, to try to find examples of all 17 symmetries on the palace's walls, floors and ceilings. It's the half-term school holiday, so I've decided to make my own pilgrimage to the south of Spain. The family, accustomed to humouring the obsessions of the mathematician in their ranks, sets off for Andalusia.

Hunting for treasure

The famous graphic designer M. C. Escher also first visited the
Alhambra in the month of October. During that first encounter, in
1922, he became entranced by the sheer variety of designs on the
palace walls. From an early age Escher had been obsessed with tiling –
covering the whole of a plane surface with non-overlapping shapes.
The first medium he experimented with was not walls or floors but
food. The Dutch often had two cold meals a day in which they ate
boterhammen, single slices of bread covered in sliced cheese or cold
meat. As a young boy Escher would try to use pieces of cheese to
completely cover his bread, leaving no gaps. In later life, cheese gave
way to angels and demons, lizards and fish.

Ever since civilizations have been building houses or constructing
roads, they have been looking, like Escher, for ways to piece together
bricks, stones or slate to cover two-dimensional surfaces and three-
dimensional spaces. The drystone walls of ancient Britain were jigsaws
of irregularly shaped stones. The brickwork looked completely random,
but the walls were sturdy enough to keep livestock in and marauders
out. Using irregular pieces certainly spared the builder the effort of
carving out regular squared-off stones. However, the lack of any pattern leads to a corresponding lack of efficiency in constructing the wall
because each new piece requires work by the builder to see how to fit
it into the construction.

Our natural passion for patterns and recognizable images soon drew
builders to create something beautiful with the space they were filling.
The Romans used small pieces of coloured tile to cover their floors
with mosaics of dolphins and senators. But the Muslims, denied the
luxury of depicting images of living things, were forced in another
direction. 'The strange thing about this Moorish decoration is the total
absence of any human or animal form. This is perhaps both a strength
and a weakness at the same time,' wrote Escher in his travel diary
during that first visit to the Alhambra.

Although the Koran itself does not explicitly prohibit pictures of the
human or animal form, many other sacred Muslim texts outlaw any
depiction of beings with a soul. One of them states that 'he who makes
images will suffer the most severe punishment on the Last Day . . . the

angels of mercy do not enter dwellings where there are such images.' Deprived of the sensuous images that other cultures used during the early second millennium, the Muslims were obliged to find more geometric ways to give expression to their artistic urges. For them, geometry and symmetry were attributes of a perfect God and suitable ways to represent his perfection in art. They identified strongly with Plato's sentiment that 'God ever geometrizes'. Indeed, the books of geometry by Euclid were some of the first Ancient Greek texts to be translated into Arabic. Armed with a sophisticated mathematical intuition, the Moorish artists began to cover their palaces in tiles of different geometric shapes and colours.

The natural world has discovered over years of evolution that only those with superior DNA achieve perfect symmetry. For the artisan too, symmetry was the ultimate test. In an age that predated the industrial mass production of perfect copies, the skill involved in repeating a design for a tile over and over again, with no flaws, was a mark of true craftsmanship. But it is in finding ever more inventive ways of combining these tiles that the artisan's mathematical prowess becomes apparent. The sultan would reward handsomely the artist who discovered a new pattern with which to decorate the walls and delight the residents.

As we make our way to the bus for the Alhambra, we walk upon a range of different pavements. The way the slabs have been arranged provides the town planner with the chance to play the same games of symmetry that the Moorish artists indulged in centuries before. I start trying to analyse which pavings have the same symmetries and which are different. Some of the patterns consist of simple squares repeated left to right, backwards and forwards, as on a chessboard. Sometimes the squares are staggered as one meets each new layer. Sometimes they are staggered in a symmetrical way, but often there is bias towards one side (Figure 23). Then you get the same thing with rectangles. One of the more intriguing pavements has rectangles zigzagging along the side of the road. I point this one out to Tomer, but my excitement is met by a long hard stare before he turns and climbs onto the bus for the Alhambra.

By the time I catch up with Tomer he's already got his Nintendo DSX out and is lost in Super Mario until we reach the Alhambra stop. I'm left to ponder the different bits of notation that I've got for the

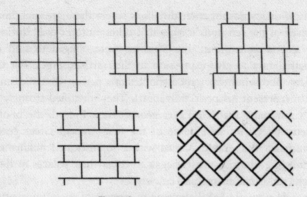

Fig. 23 Symmetry in pavements.

different symmetry groups. Included in this notation is a label for each of the different groups of symmetry: p4, cm, p4gm, and so on. The labels have been chosen in an attempt to reflect each group's specific traits. I've copied out the conventional labels for the 17 symmetries in a notebook I always carry with me. My notebook is where any flash of inspiration or idea for a problem will be jotted down – ideas can come to me in the strangest places. I've made some sketches of all the pavements we saw on the way, and I'm trying to categorize their symmetries using the notation. Either I'm being really stupid or it's incredibly bad notation, because I can't seem to sort out which name goes with which pavement. This should be trivial for me, but the whole thing is quite subtle.

Language, notation and naming are all extremely important in capturing the essence of a mathematical structure. The notation I've jotted down was created at the beginning of the twentieth century by crystallographers rather than mathematicians. The symmetries found in the floors and walls are also important for the chemist as they are related to understanding crystal structures. But the language they've devised to label these symmetries is somewhat opaque. I've also come armed with a second new notation composed recently by John Conway. It makes more mathematical sense, and as soon as I start using these labels the symmetries in the pavements become quite transparent.

As we walk into the Alhambra I'm immediately struck by the reflective power of water. It seems as though the palace is built on water. Tiny little streams run from one fountain to another. The architecture of the palace already demonstrates a vertical symmetry: the left side of the facade is perfectly reflected in the right side. Standing at one end of the pool in the Courtyard of the Myrtles, one sees another perfect copy of the building reflected horizontally in the surface of the water. This vast expanse of water stretches to the very feet of the columns of the facade, and the palace and its reflection combine to give the impression of a crystal suspended in the sky.

Some girls thrust their hands into the water, trying to touch the fish swimming there, and the symmetry is destroyed. It doesn't take much to disturb the water's calm surface and fragment the palace's mirror image. That is the message in the water: perfect symmetry is hard to obtain. The natural world knows that. The Moorish architects loved the symbolism of the fragility of the symmetry in the water. The tension between the eternal nature of God and the transience of our fragile earth was captured by the dialogue between the solid symmetry of the palace and its elusive reflection in the pool.

The palace is decorated with many images full of rotational symmetry. The Moors, like the flower and the starfish before them, discovered the power of the symmetry contained in the many-pointed star, and covered the ceilings, walls, floors and gardens of the Alhambra in stars of ever more intricate artistry. They used a trick in carving these stars which is also played by flowers. The five-petalled flower of the magnolia does not have mirror symmetry. Instead the petals are arranged such that they fall under and over each other, destroying the reflectional symmetry. It gives a spiral effect, either clockwise or anticlockwise.

The same trick was used by Escher when he put a twist on the starfish that adorns the chocolate box he designed and Conway so loves. The spiral effect is also used in many of the stars in the Alhambra. These stars are often built by interweaving several squares to realize the 8, 12 or 16 points of a star, an illustration of how the symmetries of these many-sided shapes can be broken down into symmetries of smaller shapes (Figure 24). Even Tomer can see that, despite the many different colourful and complex decorations that have been used, the symmetries of the beautiful eight-pointed star are the same as the eight rotational symmetries of a simple octagon. Both are just different manifestations of the same type of symmetry.

The walls of the palace are covered with tiles of different colours

Fig. 24 An eight-pointed star made of two interlocking squares.

arranged to make repeating patterns. Although the tiles stop where a
wall ends, the symmetry creates the impression that beyond the wall
the pattern continues to repeat itself. There is a rhythm created by the
symmetry which almost makes the walls pulsate, giving the effect of a
moving image, hinting at the infinite expanse of space. This is another
reason why the Muslim artists were drawn to symmetry: as an artistic
expression of God's infinite wisdom and majesty. Each fresh wall offers
the artist the chance to create a different, original tiling. But can one
make a science out of this art? Is there a mathematics which will reveal
ever more intricate patterns or show the limitation of what is possible?

The essence of a tiling is that it repeats itself in two directions.
Figure 25 shows the tiles on the first wall that greets visitors as they
enter the Alhambra. Eager tourists rush past me on either side, failing
to notice the images carved on the walls lining the entrance, which
almost seem to be saying 'Welcome to the palace of symmetry.' I am
struck by the extraordinarily shaped tile that the artist has discovered
which seems to fit so perfectly round the eight-pointed star to leave
no gaps.

What makes the images at the entrance to the Alhambra a regular
tiling and not a Roman mosaic or Escher cheese sandwich is that each
piece can be lifted and shifted (either up or down, left or right) and
eventually it will sit perfectly on a copy of itself. But there is more
regularity here than in just the individual movement of each piece.

Fig. 25 The wall of the entrance to the Alhambra.

I can take a copy of the whole picture, shift it horizontally or vertically, and lay it down again so that it exactly matches the original picture. This is what imbues it with a sense of the infinite. The symmetry in the wall contains a message – a programme, if you like – which stipulates exactly how the tiles will be laid out as the wall is expanded, even to the infinite reaches of the universe.

But there is more to the symmetry of this wall than simple repetition. How can we articulate what that symmetry is, though? How can we express the fact that one wall has more symmetry than another? Is it possible even to pin down precisely what we mean when we say that two walls have the same symmetry?

The reason there is more symmetry in this wall than simple repetition is that there are other ways I can pick the picture up and place it down on a shadow of itself. Instead of simply shifting it left or right, up or down, I can turn it before I lay it down. For example, if I keep the centre of one of the eight-pointed stars fixed and rotate a copy of the picture by 90° around this point, the shapes line up perfectly on top of the original picture.

I am intrigued to see what Tomer makes of the design. His initial reaction is that it hasn't got any symmetry. He is looking for lines that he can fold the image along so that the two sides of the picture match up, as one of the psychologist Rorschach's inkblot images. Immediately he can see that this isn't possible here. Intriguingly, the design that adorns the entrance to the palace is missing the version of symmetry that most people are familiar with: reflectional symmetry.

Although the eight-pointed star has mirror symmetry, the elongated T-shape that surrounds the star has a clockwise twist to it that would be reversed in a mirror. As with one's left hand, a reflection of the T gives us an entirely new shape. If I lift the piece out from the wall and turn it over, I get the reflected image. But it is impossible to place it back in the gap that has been left behind. Of course I could take out all the pieces, turn them over and place them back down around the eight-pointed star. This will produce a different (but obviously related) picture, which seems to spin in an anticlockwise pattern.

The secret of articulating the symmetry in this wall is to imagine all the ways you can lift the image and place it back down in an outline of the wall. It helps to imagine a ghostly form of the object being left behind as the object itself is moved and then replaced. Following in a long line of artists who have come to the Alhambra, including Escher, Tomer and I have brought sketchbooks in which to draw the walls. I actually find it remarkably difficult to sketch the complicated template used by the Moorish artist, and begin to feel a certain wonder at the way it perfectly covers the wall.

I look up to see it hasn't taken long for Tomer to have dropped his sketchbook and whip out his Nintendo DS. But when I berate him for escaping into Super Mario Karts and not appreciating the beauty around him, he shows me the screen. What he's been doing is drawing the designs onto the screen. Not only that, there is a function that allows him to spin what is on the screen. He shows me that if you fix a point at which the strange elongated T-shapes meet in a sort of swastika, then you can spin the picture round this point by 90°. Sure enough, there on the screen he animates for me the tiling spinning around this point and lining up again so that it looks exactly the same.

The Nintendo actually captures something my simple sketches can't – the fact that the symmetry of the tiling is about movement. It is about the variety of things I can do to the picture. Although the picture ends up looking like it did before it was moved, the essence of each new symmetry is the intermediate motion that gets back to the original picture – something the animation on Tomer's Nintendo has captured but my simple static sketch only hints at.

Tomer's drawing on the Nintendo has captured another symmetry which is distinct from the symmetry that spins the picture around the centre of the eight-pointed star. Although both are rotations of 90°, there is a genuine difference between the two because of their different effect on the picture. Tomer wanders off smugly into the palace.

Following sheepishly, I find myself in a room covered with tiles of different shapes to those at the entrance. A guide is pontificating about the wonders of the Arab mathematicians: 'To create a square is easy, but without the aid of the computer the Arab mathematicians created here not just eight-pointed stars but as many as 16 points on a star.' They certainly excelled in their mathematics, but it is slightly fanciful to imply that one needs a computer to generate such stars. When the guide then claims the Arabs also invented zero, I have to hold myself back from launching in with an explanation that the Indians discovered zero. The Arabs were just good messengers bringing the idea from the East to the West along the silk routes.

On the ceiling is another style that the artists in the Alhambra particularly enjoyed using (Figure 26). They often carved the wood or laid the tiles to create lines that look as though they run under and over each other like a tangled knot. The walls and ceilings look like a basket rendered in wood or stucco. It cleverly tricks the eye into providing an extra dimension to the experience of looking at the two-dimensional wall and again destroys any possibility of simple

Fig. 26 The ceiling in the first room of the Alhambra.

reflectional symmetry. In any mirror image of the picture, the order in which the lines run under and over each other will be reversed. I am intrigued to see whether Tomer sees this as a different sort of symmetrical pattern to the wall that greeted us at the entrance. Or are their symmetries the same?

'But they're completely different pictures,' Tomer declares. My partner Shani comes over. She's an artist. What does her increased sensitivity to the artistic side of symmetry make of the two pictures? She concurs with Tomer: 'Not the same.' A closer look shows that, as with the design at the entrance to the Alhambra, you can fix a point at the centre of the octagon and spin it through 90°, and the ceiling pattern will look just like it did before the spin.

'Look – there's another place you can spin it.' I point to the centre of the tiny square at the heart of each group of four octagons. The square is formed by the two white strands weaving in and out of each other. The pattern made by the strands actually means that you have to spin the picture by 180° before it sits back perfectly within its outline. 'That's starting to make my brain go all fuzzy.' Tomer pretends to faint to the floor. At first sight this ceiling pattern looks quite different to the tiles at the entrance to the Alhambra. Even with my more sophisticated perspective, I'm not actually sure whether these are the same symmetries or not.

But when I go back outside and check, I see that there is also a point around which the wall tiles can be spun by half a turn. The point around which you have to spin the picture is the halfway point between one eight-pointed star and its closest neighbour. Just as with the pattern on the ceiling, you have to spin the wall tiling a full 180° before the images line up again. This is quite embarrassing – symmetry is my speciality, yet I'd missed this symmetry on first viewing.

An even bigger challenge is to work out whether these two different designs have the same symmetry type or not. It is one thing to count the number of symmetries in each pattern. But how does one talk about the totality of symmetries? Does it make sense as a concept in its own right? In the nineteenth century, mathematicians would eventually devise a language to articulate the fact that these two patterns, one at the entrance to the Alhambra and one on the ceiling, do in fact have the same group of symmetries, meaning that the rotations have the same effect on the objects they are spinning. Even without this language, we can illustrate that the symmetries of the patterns are related by superimposing one on the other. If we twist the ceiling pattern through 45° and place it over the entrance wall tiling in such a way that half the octagons sit on the eight-pointed stars and the other half have their centres where the swastikas are, we can see how they fit (Figure 27).

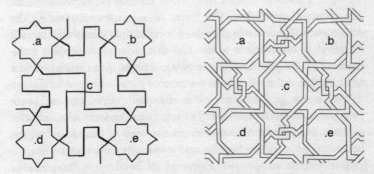

Fig. 27 Lining up two patterns reveals their symmetries to be the same.

Once the pictures have been so aligned, any symmetry of one picture directly translates into a symmetry of the other. It's rather like the way you can place a dodecahedron inside an icosahedron (or vice versa; see Figure 22, page 58) and see their symmetries match up. But once they were equipped with a language for symmetry, mathematicians found that they could describe such alignments without having to look for any physical resemblance.

The most powerful result of speaking this new language was that it would eventually allow mathematicians to prove that there are no more than 17 different types of tiling symmetry. Any pattern that is repeated both vertically and horizontally must fit into one of these 17 classes. On the walls of the Alhambra Tomer and I have so far identified just one of these 17, called 442 in Conway's notation (nothing to do with football, I should add). The two 4's indicate that in this symmetry there are two different sorts of 90° rotation – something which is not so clear from the picture in the ceiling. The 2 refers to the half-turn we missed on the tiles at the entrance.

Tomer is getting restless. 'Come on, dad. We'll never get out of the palace at this rate.' Our task now is to sniff out as many of the other

16 symmetries as we can. My illustration skills can't keep pace with the onslaught of images that greet us at every turn, so I resort to the digital camera.

Triangles and hexagons, gyrations and miracles

The group of symmetries that we've found at the entrance to the Alhambra is in fact a subgroup of the very simple symmetries created by the plain undecorated square tiles that cover most people's bathroom walls. By putting more elaborate designs into the picture, the artist has killed the reflectional symmetries of the simple square tiling. As well as the simple square tiling, there are two other very simple patterns on the wall that are full of symmetry, variants of which start to make an appearance as one enters the heart of the Alhambra: the hexagonal lattice of the beehive, and a wall covered with triangles.

Tomer spots an interesting network of interlocking three-headed arrows (Figure 28). There is a reflectional symmetry here as well as an obvious rotational symmetry around the point labelled A at the centre of the arrow. But hidden inside the way these arrows are put together is a less obvious rotational symmetry. The point B, where three of the arrowheads meet, is another point about which the picture can be spun through a third of a turn. The rotation about B has a different quality to the rotation about A because the point B does not lie on

Fig. 28 Interlocking three-headed arrows. A and B are two points around which one can rotate the picture by a third of a turn. Point B does not lie on a line of reflection. The rotation around such a point is called a gyration.

any line of reflection. Conway's name for this sort of symmetry is a gyration. The complete group of symmetries Conway denotes by 3*3. Conway uses a star to indicate some reflectional symmetry. Any gyrations in the symmetry are indicated by numbers before the star, one number for each different gyration.

By now I've found nine of the 17 different sorts of symmetries. On a column I spot a pattern of leaf-shaped tiles which starts to push one's idea of symmetry beyond conventional reflections and rotations (Figure 29). There are some simple reflections and translational symmetries that one can perform on this tiling. For example, ignoring the two shades of tile, I can pick the pattern up and shift it diagonally so that point A ends up at C and B ends up at D. The white leaf ends up sitting on top of its black neighbour. But there is another symmetrical move that I can make: pick the picture up, reflect it, and then shift the image up and along. Then, for example, A ends up at D and B ends up at C. This sort of symmetry is often much harder to spot. You can't do this move simply by reflecting in a line through the picture. It is what some call a glide symmetry, but Conway prefers to call 'a miracle' or 'miraculous crossing'. For him, this strange symmetry produces a mirror image of the motif without the presence of a mirror. The name 'miracle' is as much about missing mirrors as it is about expressing a sense of wonder at the discovery of such a strange symmetry.

I show this to Tomer, but he takes some convincing that this really

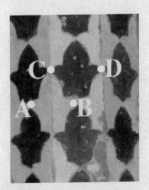

Fig. 29 This tiling illustrates a new sort of symmetry, called a glide or miracle. In this symmetry the tiles are reflected then shifted, so that A moves to D and B moves to C.

is a different symmetry from simply shifting the picture up and to the left. The point is that all these symmetries leave the picture looking essentially the same, so it is sometimes hard to recognize what is a genuinely different symmetry. Putting labels on the picture actually helps us to see this. With the miracle symmetry, points A and B get transformed to points D and C, whereas the shift symmetry simply moves A and B to C and D without any reversal. This labelling is actually at the beginnings of a language that will help capture the underlying symmetry of the pictures we are looking at. Pictures are starting to give way to letters and language.

If I ignore trying to match up the colours of the leaf tiles on this column, then I've managed to add a tenth symmetry to my list. It's called *×, the star referring to the simple reflection down the centre of any tile. The × represents the miracle or glide symmetry where I reflect then shift. However, if I do take the colours into consideration and insist that white tiles go to white tiles, and black to black, then I get a different group of symmetries. My miracle or glide symmetry doesn't work any more because it changes white tiles to black tiles. The only symmetries are reflections in a line through the white tiles or a line through the black tiles. This symmetry type is called ** to indicate two different sorts of reflection.

So I pick up two groups of symmetries for the price of one. That

Fig. 30 A floor in the Alhambra illustrating another example of symmetry group 442.

gets me to 11 out of 17. Again and again I think I've come across a design with a new symmetry group only to find that once again it's 442, the symmetry group that opened our treasure hunt. Even the floor has it (Figure 30). It looks simpler than the two I found earlier at the entrance to the palace, but again, I can align the images one on top of the other to unmask them as examples of the same group of symmetries. Ultimately, mathematicians would find a more abstract language in which to talk about why two groups of symmetries are the same and which would extend this trick of aligning pictures. I still have my task of teasing the last six groups of symmetries out of the walls, floors and ceilings of the Alhambra.

I first came to the Alhambra 20 years ago, when I was an undergraduate. I spent the summer with a friend inter-railing round Europe, surviving on tins of tuna and sleeping in railway stations. Perhaps it was because of sheer exhaustion, but when I eventually reached the Alhambra I never played the symmetry hunting game. Perhaps I was trying not to look the geeky mathematician in front of my travelling companion.

A few of my colleagues obsessively collect examples of the 17 different symmetries wherever they travel: a friend's shirt will suddenly elicit a rush of excitement: 'Ooh, your shirt's got two miracles on it. Wait there while I get my camera!' Or, in the Alhambra, 'There's a gyration. Wow, that's a great *632!' In fact, the Alhambra has always been something of a challenge to symmetry nuts: there has been a lot of debate about whether you really can find all 17 symmetries inside the palace walls.

It took Escher a second visit to the Alhambra for the Moorish designs to make the deep impression on his work that now is the trademark of his style. He first visited Granada in 1922, but much of his artistic output at this time consisted of very three-dimensional graphical representations of Italy, where he met and settled with his wife. A favourite subject was the Amalfi coastline, with its harbours and its villages hanging from the cliff-faces. These images are in stark contrast to the art he produced after his next visit to Granada. He made this second trip in May 1936, during the shifting political climate of pre-war Europe. The Escher family had left Rome in the summer of 1935, frightened by the increasingly ominous atmosphere that was sweeping the country. A winter in Switzerland made the family hanker for the sun. So Escher put together a proposal to an Italian shipping company, Adria, based in Fiume:

> I suggested they take me along as a free passenger. In exchange I would give them four copies of each of twelve graphic prints; these then could be used for advertising purposes in the tourist trade. To my great surprise, they agreed.

The ship *Rossini* took Escher and his wife to the south of Spain, from where they travelled to the scene of his visit in 1922. No words can convey the impression that the Alhambra made on Escher better than one of his most famous woodcuts, *Metamorphosis*, made a year after that second visit (Figure 31). On the left, the coastal town of Atrani is depicted. But as the image evolves across the print, the three-dimensional buildings of the town morph first into cubes and then into a hexagonal tiling of two-dimensional Chinese boys.

Fig. 31 *Metamorphosis* by M. C. Escher.

This two-dimensional space began to take over Escher's world. He filled his notebooks with all the two-dimensional designs he saw around the palace, and when he returned to his native Holland his artistic output began to undergo a similar metamorphosis:

> In Switzerland, Belgium and Holland I found the outward appearance of the landscape and architecture less striking than that which is to be seen particularly in the southern part of Italy. Thus I felt compelled to withdraw from the more or less direct and true-to-nature depiction of my surroundings. No doubt this circumstance was in a high degree responsible for bringing my inner visions into being.

But it was not only that he was surrounded by the flat two-dimensional landscape of the lowlands. The sweeping tide of fascism was deeply disturbing for Escher, and this new inner world was escapism for him: 'The fact that from 1938 I concentrated on the interpretation of personal ideas was primarily the result of my departure from Italy.' Escher commented in the travel diary that he kept on his trip to Granada 'There are hardly any foreigners. We're being gaped at like creatures from another planet.' Nowadays, things couldn't be more different. The palace now restricts the numbers entering each day. Today I have been overtaken by countless tour groups being rushed through the palace with barely a chance to snap a simple reflection of the Comares tower in the pools of the Courtyard of the Myrtles.

Actually I seem to have lost Tomer. He's been caught up by the tide of tourists and whisked through to the Courtyard of the Lions. I eventually find him in the Hall of the Abencerrajes, just off the courtyard. Although it is not one of the missing symmetries I am looking for, the room has a breathtaking ceiling full of interesting geometry. A magnificent eight-pointed star spans the roof of the hall, and carved into the ceiling are thousands of little alcoves. The effect is of a ceiling covered in stalactites. It features in a lot of Muslim architecture, and recalls Muhammad's visitation by the Archangel Gabriel in the famous cave of Hira. The mathematical skill that helped create the perfect arrangement of 5,416 pieces in the eight-pointed-star ceiling is quite staggering.

One of the guides is telling the gory story of how this room is named after 36 members of the Abencerrajes family who were hacked down in this room on the command of the sultan. The guide is pointing to a red streak running through the marble floor. 'The blood of the Abencerrajes family still stains these floors.' I can't bring myself to break the illusion and tell Tomer that it's more likely to be the oxidation from the pipes bringing the water to the fountain at the heart of the room. Sometimes myth is more fun than science.

But I can't resist telling him about another of the Moors' fantastic scientific achievements, built into the Courtyard of the Lions. The windows and pillars throughout the palace have been deliberately constructed so as to make the Alhambra a huge sundial. As the sunlight streams in through the windows, the alignment of the courtyard ensures that the shadows cast by the pillars turn throughout the day like the hands on a clock. The positions of the windows even ensure that more sunlight streams into the courtyard when the sun is low in the winter, as it is today, warming the chill Andalusian air. In the summer, although the sun is higher in the sky, the arrangement of the windows lets less light into the courtyard, keeping it as cool as possible for the sultan and his harem.

I manage to pick up one more symmetry from a rather beautiful pattern running round the top of one of these columns in the Courtyard of the Lions, taking me up to 14 out of 17. But by this time Tomer is almost literally climbing the walls. 'Can we go now? I think I got the idea.' But I am slightly obsessive about my quest and won't be content until my tally is complete. I've trailed all through the palace, yet it seems that I'm missing three symmetry groups. 'Come on, dad! Can we go to the shop now?' 'All right, all right.' But I'm already planning to return tomorrow.

On the way back to the hotel we walk on the pavements we saw in the morning. My eyes and brain are now hypersensitive to all the symmetry. I feel like a bee flying round the garden, able to pick up only the outlines of hexagons and five-pointed flowers. I suddenly realize that one of the missing symmetries is staring me in the face on the urban floor of modern Granada (Figure 32). It doesn't have any reflectional symmetry or rotational symmetry. But hidden in the pattern are two strange glide symmetries – a double miracle.

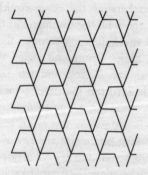

Fig. 32 A pavement in Granada containing a double miracle.

Back at the hotel, I pore over the images I've collected on the digital camera to see whether I really did get 14 different symmetry groups.

Despite having the mathematical techniques at my fingertips, it is remarkably challenging to determine whether one of the missing symmetries is among the photos I've collected. Eventually the batteries in the camera die on me and I'm forced to retire to bed. We're meant to be heading off to less mathematical climes tomorrow, but before we leave I'm determined to see whether I can track down the last three symmetries in the Alhambra. I've spotted this double miracle symmetry on the pavement, so surely I can find it in the palace.

One of the television programmes that made a big impact on me during my adolescent search for all things mathematical was Jacob Bronowski's *The Ascent of Man*. As I drift off to sleep in my hotel room I have a strong recollection of a scene in the programme where Bronowski is sitting in the Alhambra talking about symmetry. I remember him being in the Harem, talking about how the walls are covered in sexy symmetry rather than pictures of sexy women. It was a real treasure trove of different symmetries, filling every available space with patterns. And I have this strong image of one of the most beautiful symmetries, made simply out of triangles, but triangles with a subtle spin on them which destroys the reflectional symmetry. This would give me one of my missing three symmetries. But I don't remember seeing anything like that in the palace today. Did I miss a room because Tomer had dragged me off to the shop? My night is awfully disturbed. Every dream seems to fragment into a hexagonal lattice or zigzags of rectangles. Getting up early, leaving the others to sleep on, I head back up to the palace.

Hidden symmetries

Every time I prove a new theorem, it is rather like constructing a new part in the palace of mathematics. But after I've finished the proof, I anxiously survey what I have achieved, revisiting the mathematical structure I've built to make sure that it won't collapse. Yesterday I went round the Alhambra in a positive, butterfly collecting mode, swishing my net this way and that, gathering whatever symmetries lay in the path of the net. Today I'm using a more critical eye, questioning everything to see if it will yield my three missing symmetries.

Scouring the walls, I notice lots of things that didn't register yesterday. It is extraordinary how the brain can take in so much data and no more. But I still can't quite root out the patterns I need to complete my set of 17 wall symmetries. I come all the way round to the Courtyard of the Lions again, with its dozen lions holding up the fountain. And there on the floor is one of the patterns I need (Figure 33). It's the same symmetry as the one I found on the pavement yesterday: a double miracle. But unlike yesterday's, where some rotational symmetry was destroyed in the tiles by their strange shape, the colours work in my favour here. It's a zigzag of rectangles alternating in colour between white and green. Without the colours you get a rotation which maps one row of diagonal tiles onto the lower row. But now the colours mess that up, leaving a double miracle.

Fig. 33 A double miracle on the floor of the Alhambra.

Now I've just got two symmetries to go. At the other end of the Courtyard of the Lions I strike lucky again (well, almost). Today I'm more sensitive to the effect of colour. On the wall on the other side of the courtyard I pick up something I missed yesterday. It's essentially six triangles arranged in a hexagon, the colours of the triangles alternating round the hexagon between red and yellow (Figure 34). By introducing the colours, the craftsman has changed the rotation of a sixth of a turn to one of a third of a turn, because to move the pattern round to the same configuration I have to position a yellow triangle on a yellow triangle. The only trouble is that, although the designer has cleverly varied the colours of the triangles, he's also introduced some annoying blue tiles that ought to be black. By pretending that the blue tiles are black, I'm able to add *333 to my collection, although it's a fudge. The artist had the right idea but just missed perfection – from my mathematical point of view.

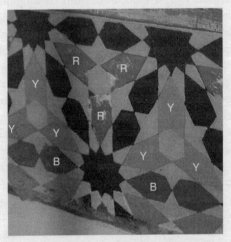

Fig. 34 A wall in the Alhambra whose symmetry group is almost *333. Y, R and B denote yellow, red and blue tiles.

Escher too was particularly sensitive to the use of colours in the Moorish designs. As he got more and more drawn to studying the mathematics behind the patterns, his increased artistic sensitivity to the importance of colour led him to introduce a new structure in the mathematics of symmetry that scientists had missed. In addition to the 17 symmetry groups, there is an extra range of symmetries based on being allowed to move pieces and swap colours around. It has been proved that, if you include permutations of two colours, there are an additional 46 symmetry groups. It is Escher we can thank for this new mathematical perspective on the walls of the Alhambra.

I've still got one more symmetry left to find. I'm after the twisted three-pointed stars that I'm sure are somewhere in the palace and would give me 632 which is the symmetry group that I'm missing. But I've almost reached the exit. I don't quite get it. Had Bronowski been filmed somewhere else? There are also great patterns in the alcázar in Seville, but I'm convinced that the footage was shot in the Alhambra.

Then I spot it. A barrier stops visitors from climbing through a small entrance. Beyond the barrier I can see a gallery overlooking something down in the lower levels of the palace. I take a quick look round. There's no one to see me, so I vault the barrier. It's not exactly Indiana Jones, but I get a slight buzz from crossing into forbidden territory. I'm someone who follows the rules – for most of the time. It's my mathematical upbringing. Mathematics very quickly begins to collapse if one strays outside the logical boundaries permitted by the subject.

The only computer game I ever got hooked on was *Prince of Persia*. I played it after hours with the secretary of the department when I was visiting Israel as a post-doc. She was fantastic at fighting (she'd been in the Israeli army after all) and I did the logic bits. Creeping around the darkened inner sanctum of the harem of the Alhambra feels like the moment we'd find our way to the next level in the game.

I look over the balcony, and there it is: the backdrop to Bronowski's piece in *The Ascent of Man*, full of symmetry. And there too is the pattern I had in my mind the night before (Figure 35). It was one of the patterns that had inspired Escher. The triangles seem to shimmer in the Andalusian sun. Their curves create a sensuous background to

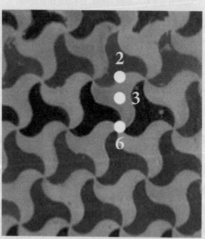

Fig. 35 A wall in the Harem with group of symmetries 632. Each point marks a place around which you can rotate the picture. The numbers indicate how many repetitions of each rotation it takes to return the tiles to their original positions.

a room that was the emotional and sexual heart of the palace. It was from the balcony where I am now standing that the sultan would look down on the women in the Harem reclining naked after their bath and select his companion for the night, sending her down an apple as a signal that she'd been chosen. The sensuous triangles on the walls of the Harem contrast with the stark squares of the formal entrance, where the interlocking pieces look almost like barbed wire protecting the building from unwanted visitors. Looking down from the balcony is like staring into not a well filled with water, but a vast pool of symmetry. Here the artists covered every available space with as many symmetrical games as they could conjure up.

What's quite special about this configuration is that by putting a spin on the three-pointed triangles, the artist has lost all reflectional symmetry. This is what inspired Escher when he twisted the starfish on his chocolate box. Yet the picture is full of different sorts of rotational symmetry. Ignoring the colours, one can spin the picture through a third of a turn about the centre of each tile (the point marked 3 in Figure 35). There is also a spin of a sixth of a turn (about the point marked 6) where the ends of the tile meet each other – remember that we are ignoring colours. And finally, a slightly more subtle rotation of half a turn is hiding inside the picture. If you fix a point (marked 2) halfway along an edge, you can actually make a half-spin which takes the tiles and lays them back down perfectly on top of the outline.

Still buzzing with the thrill of finding my missing symmetry, I quietly slip back along the corridor leading to the barrier I'd leapt over. As I emerge, a startled group of Dutch tourists look rather outraged at my rule breaking. Striding past them, I head back for a last look around the Courtyard of the Lions. But then my eye is caught by something else I've missed – twice, now – on my journey through the palace. Half-hidden behind a wooden screen is another pattern I haven't seen before (Figure 36). In this pattern there are no lines of reflection. But look at it carefully, and you should see points about which you can spin through a sixth, a third and a half-turn. Amazingly, it's got exactly the same symmetry as the three-pointed stars I've just uncovered in the Harem. Have the authorities decided that this symmetry type is just too racy for untutored eyes, and tried to keep it out of public view?

Fig. 36 Another wall in the Alhambra with the group of symmetries 632.

With a little bit of creative repainting, my treasure hunt has yielded all 17 different wallpaper groups. But how can I be sure that there isn't an 18th one waiting out there to be discovered? Certainly one can play lots of games with the actual shapes that are used as tiles. Escher's bats, angels, lizards, fish, birds, butterflies, beetles, crabs, bees, frogs, griffins and seahorses illustrate the infinite variety of forms that can be used. But hiding behind each tiling pattern is one of only 17 different varieties of symmetry that is possible on a two-dimensional surface.

It would be another five hundred years before mathematicians proved for certain that the medieval Moorish artists would never have been able to squeeze an 18th type of symmetry out of the tiles on the walls. As we shall see later in our story, it is a proof that depends on mastering group theory, the nineteenth-century language for capturing the subtleties of symmetry. It is thanks to the unique power of mathematics that we can say categorically that there can never be an 18th pattern, despite highly creative attempts to show otherwise.

Escher explained that it was only when his brother, a geologist, referred him to a series of academic papers on the mathematics of symmetry that he gained a complete understanding of what he saw around him. He described his first impression of trying to come to terms with the onslaught of ideas:

I saw a high wall and as I had a premonition of an enigma, something that might be hidden behind the wall, I climbed over with some difficulty. However, on the other side I landed in a wilderness and I had to cut my way through with a great effort until I came to the open gate, the open gate of mathematics.

This experience is shared by every mathematician who has entered this magical world. Escher continued:

From there, well-trodden paths lead in every direction, and since then I have often spent time there. Sometimes I think I have covered the whole area, I think I have trodden all the paths and admired all the views, and then I suddenly discover a new path and experience fresh delights.

Escher had not been very interested in mathematics at school. 'I was extremely poor at arithmetic and algebra,' he said, 'because I had great difficulty with the abstractions of numbers and letters. But our path through life can take strange turns.' But ultimately, to find a way to capture the pictures that adorn the Alhambra, mathematicians realized that they would have to translate them into the language of algebra and letters and enter the abstract world of the mind.

The language that mathematicians created to navigate the world of symmetry had its genesis in a completely different problem. While the Moors in Spain were painting symmetry on the walls of the Alhambra, Arab mathematicians in Baghdad had been making progress on the seemingly unrelated problem of how to solve equations. Neither could have predicted that, over several centuries of mathematical development, these two major themes in the mathematical opus should gradually interweave until they became inextricably linked. The new language would allow mathematicians to go beyond the walls of the Alhambra and understand the limitations of symmetry throughout the whole mathematical palace, in three-, four- or even higher-dimensional space.

It was said that as the last Muslim sultan fled Granada in 1492 when the Christians took the city, he turned to take one last look at the Alhambra and wept. His mother chastised him with these harsh words: 'Do not weep like a woman for what you could not defend like a man.' I can understand the sultan's distress at leaving behind something so beautiful.

November: Tribal Gathering

> Beware of mathematicians, and all those who make empty prophecies. The danger already exists that the mathematicians have made a covenant with the devil to darken the spirit and to confine man in the bonds of Hell.
>
> ST AUGUSTINE, *De Genesi ad Litteram*

1 November, Okinawa

Science is about discovery, but it is also about communication. An idea can hardly be said to exist if you do not awaken that same idea in someone else. That is why conferences are an important part of giving life to an idea. It is one of the most exciting parts of the job, because you get to perform the mathematics to an interested audience. A proof is like a piece of theatre or music, with moments of high drama where some major shift takes the audience into a new realm. I'm on my way to a meeting in Japan to share my perspective on the world of symmetry.

I like journeys. There is something about movement which helps my thought process. Trains are my favourite form of transport for inspiration. Staring out of a train window, letting images flood my vision at 125 miles an hour, I find the perfect stimulant for mathematical creation. My DPhil thesis was the result of a flash of inspiration one afternoon on the train from Reading to Oxford (granted, not a train that hits 125 miles an hour).

As I take my seat on the plane, the man next to me starts grinning inanely. Not a good sign. It's a 13-hour flight, and I normally try to ward off any conversation until the last five minutes as we're coming in to land. So I hide behind my yellow pad and start scribbling.

'What do you do, then?' he asks. I hope the discovery that I'm a mathematician will frighten him off. Most people's faces freeze when they find this out. Then they mutter about how bad they were at maths at school and always feel the need to tell me what grade they got in their maths O level or GCSE. 'I quite like math.' He's dropped the 's', so he's American and he clearly isn't going to spend the rest of the flight in silence. He presses me further: 'So what sort of thing do you do?'

The great German mathematician David Hilbert declared in his famous lecture in 1900 to the International Congress of Mathematicians that 'A mathematical theory is not to be considered complete until you have made it so clear that you can explain it to the first man whom you meet on the street.' So I can't resist trying Hilbert's maxim out on my fellow passenger. 'Do you want the one-minute, five-minute or 13-hour version?'

Frenzied and innumerable

For me, communication goes beyond just telling stories to those who also speak this secret language of mathematics. When I was a student in Oxford I used to spend a lot of my time trying to explain to people who were studying other subjects why I was so passionate about mathematics. I fell in with a set of people doing an eclectic mix of studies: Persian and Arabic, Philosophy, Politics and Economics, Literary Theory. Increasingly I found myself trying to explain at parties or late-night sessions in someone's room why I thought doing maths was as exciting as deconstructing Henry James.

People vaguely get the idea of what's involved in being an ecologist studying the Amazon, a physiologist investigating medicine in space, or a marine biologist scouring the seabed in his submarine. But what on earth (or rather not on earth) a mathematician gets up to is still a complete mystery to most people. I would try to give people a small glimpse into my world and show them why I find it as magical as the Amazon, outer space or the bottom of the ocean.

I realized that I'd made some progress one day in the library. Nicki, who did English, came up to me and laid a book down on top of my maths. 'This sounds a bit like what you keep on saying mathematics is all about,' she said. She was pointing to a quote from Jorge Luis Borges. In one of his many short stories, Borges had invented a Chinese Encyclopedia according to which animals were classified as follows:

  (a) belonging to the Emperor;
  (b) embalmed;
  (c) tame;
  (d) sucking pigs;
  (e) sirens;
  (f) fabulous;
  (g) stray dogs;
  (h) included in the present classification;
  (i) frenzied;
  (j) innumerable;
  (k) drawn with a very fine camel hair brush;
  (l) et cetera;
  (m) having just broken the water pitcher;
  (n) that from a long way off look like flies.

The quote perfectly encapsulates the centuries-old pursuit of symmetry. Each new step in the mathematician's journey would add crazier categories of symmetrical beasts to the list. The Greeks discovered the five Platonic solids, whose symmetries make them perfect objects for dice. The artists who decorated the Alhambra tiled the walls of the palace with 17 different types of symmetry. In the twentieth century, Conway's Atlas documents an ever wilder and eclectic selection of symmetrical objects, culminating with the Monster, whose description sounds as bizarre as Borges' animals 'that from a long way off look like flies'. As I continued my own journey, trying to classify what symmetrical beasts can be built from the animals in Conway's Atlas, I found the quote more and more apposite, so much so that I decided to open my doctoral thesis with it.

I also fell in love with Borges. He is a mathematician's writer. His short stories are like mathematical proofs, delicately constructed and with ideas laced together effortlessly. Each step is taken with precision and watertight logic, yet the narrative is full of surprising twists and turns.

Part of Borges' Encyclopedia is particularly relevant to the project I'm trying to tackle at the moment. I've taken the simplest animal in the Atlas of symmetry, the rotations of a prime-sided shape, and I'm trying to classify the symmetrical shapes that can be built from this basic building block. Establishing the range of possibilities though is extremely complex, and most mathematicians had relegated these symmetrical objects to categories (i) frenzied and (j) innumerable of Borges' classification. My mathematical ancestor Philip Hall – my supervisor's supervisor's supervisor – declared that 'the astonishing multiplicity and variety of these groups is one of the main difficulties which beset the advance of finite group theory'.

The project I am battling with at the moment is how to rescue these objects from category (j) of Borges' classification: to enumerate the groups of symmetry that can be built from the rotations of a triangle. The Greeks had identified five Platonic solids. The Moors had painted 17 different symmetries on the walls of their palaces. Can I find a way to count how many different objects there are with $3^2 = 9$ symmetries, with $3^3 = 27$ symmetries . . . with 3^{10} different symmetries? I may not know exactly what they all look like, but my hope is that there is some way to count how many there are. Perhaps I can spot some pattern to the way the number of objects grows as I add another triangle each time.

Although most of the things I'm looking for are rather abstract and live in higher dimensions, the symmetrical objects built from the symmetries of two equilateral triangles can still be seen in two and three dimensions. The shapes will have $3 \times 3 = 9$ symmetries. It turns out that there are two genuinely different symmetrical objects with nine symmetries. The shapes may have the same number of symmetries, but these nine symmetries behave very differently in each object.

The first of these objects is the group of rotations of a nine-sided regular polygon, a nonagon (Figure 37). There are nine different rotations of a nine-sided coin which leave the coin inside an outline drawn around it, including leaving the coin where it is.

The second object with nine symmetries can be built by taking a black triangle and a white one and pinning them together, one on top

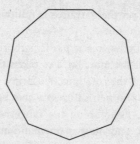

Fig. 37 The nonagon has nine rotational symmetries.

of the other, so they look a little like a combination lock with two triangular wheels (Figure 38). I used a similar object to analyse the throw of the dice in the Game of Ur from the British Museum. The symmetries of this shape are got by spinning the two triangles independently. Each individual symmetry is one of the magic trick moves that leaves the two-wheeled combination lock looking unchanged. To keep track of how many different moves there are, it helps to put numbers on the sides of the triangles, as on a real combination lock.

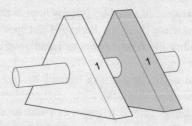

Fig. 38 A combination lock with nine symmetries.

Anyone who has forgotten the combination for such a lock has probably contemplated going systematically through all the numbers to rediscover it. I can use the same trick to analyse the symmetries of this object. For example, I can leave the white triangle stationary and spin the black one round by a third of a turn. That leaves the numbers 1 and 2 showing, which I denote by the notation (1, 2). I can also spin the black triangle through two-thirds of a turn, to leave (1, 3). With this notation I can easily keep track of all the different symmetries. There are nine different moves I can make:

(1, 2), (1, 3), (2, 1), (2, 2), (2, 3), (3, 1), (3, 2), (3, 3), (1, 1)

The last of these is the magic trick move where I just leave the triangles as they are.

The nine different permutations of the combination lock correspond to the nine symmetries of the object. The locks that are used on a briefcase, for example, generally have three wheels with ten numbers on each wheel. So there are $10 \times 10 \times 10 = 1,000$ symmetries you have to try before you've checked every combination, which is why this lock is reasonably secure from an opportunistic attack.

Starting to emerge in the analysis here is a language in which symmetries can be expressed as numbers. The numbers make it much easier to keep track of how many different symmetries there are. Each pair of numbers actually identifies exactly what the symmetry move is. Translating geometric moves into numbers would eventually allow mathematicians to begin to decode the complete book of symmetry.

One of the key issues is determining whether one has built a truly new group of symmetries or has merely found a previously known group in a new guise. In the Alhambra I kept taking photos of wildly different looking designs, convinced that I'd found a new symmetry to add to my list, only to discover that the wall had the same symmetry as something I'd recorded earlier. So how can I be sure that the combination lock is a genuinely different symmetrical object to the nine-sided polygon? Could they actually be different manifestations of the same group of symmetries? After all, they both have exactly nine different symmetries. This is one of the difficulties facing anyone who tackles this subject: two objects can look very different yet have the same underlying symmetries.

If I repeat any of the symmetrical moves of the combination lock, after three moves the lock will be back to how the triangles were set when I started. Take the symmetry that moves the white triangle forward a third of a turn and the black triangle back a third (Figure 39). So after this move, the lock has gone from (1, 1) to (2, 3). Now

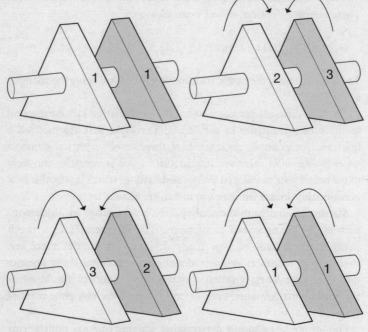

Fig. 39 The effect of repeating symmetry (2, 3).

repeat this same move. The lock moves on to $(3, 2)$. Repeat the move once more and $(1, 1)$ appears again. Whatever symmetry move you choose to implement, repeat it three times and the combination lock will have returned to its original position.

Now let's look at the nine-sided polygon. Here there are rotations which require nine repetitions before the figure comes back to its original position. For example, turning the polygon through one-ninth of a whole rotation clearly requires doing nine times to get the polygon back to its starting position (Figure 40). So the two groups of symmetries are not the same. This explanation illustrates an important lesson in the theory of symmetry, which came to be fully understood early in the nineteenth century: that the nature of the underlying symmetry of an object starts to reveal itself only when you begin to explore what happens when you combine symmetrical moves.

The rotational symmetries of the nine-sided polygon and the

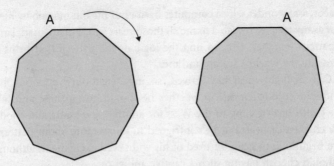

Fig. 40 Repeating this rotation nine times brings A back to its starting point.

combination lock with two triangular wheels are the only two symmetry groups with nine symmetries. But as you add more triangles and ask, for example, how many objects are there with $3 \times 3 \times 3 = 3^3 = 27$ symmetries, then more exciting things happen. Two of the objects with 27 symmetries are built in a very similar fashion to the previous two examples. A regular polygon with $3^3 = 27$ sides has 27 rotational symmetries. Or you could build a combination lock with three different triangular wheels whose symmetries correspond to the $3^3 = 27$ different ways in which you can spin these triangles. But in addition to these two, mathematicians discovered that there are another three symmetrical objects that each have $3^3 = 27$ different symmetries, making five objects in total.

As I add more and more triangles, the number of possible symmetrical objects goes up. There are 15 objects made from the symmetries of four triangles, and 67 made from the symmetries of five triangles. But it is a complete mystery how many symmetrical objects you can make from the symmetries of ten triangles. I am trying to find a way to predict how the numbers of objects grow as I add more triangles.

The pattern hunter

The challenge of trying to find patterns in the way these numbers evolve as I add more triangles goes to the heart of what it means to me to be a mathematician. Many of my friends have the impression that I'm sitting in my office doing long division to a lot of decimal places, and wonder why a computer hasn't put me out of a job by now. But as my teacher revealed to me all those years ago, a mathematician is a pattern searcher. I try to find the logic or the pattern that helps to generate the world I see around me.

Our in-flight meal has arrived, so my neighbour's attention has drifted a little to engage in whether he should go Japanese and take the bento box or cling to the West for another few hours and choose the chicken or beef. But he's intrigued to know more about patterns. I'm beginning to get a bit tired of his wide-eyed excitement, although I should cherish having such a captive audience.

To give myself a breather, I set him a little challenge that I hope will keep him going for the next few hours. What's the next number in this sequence:

13, 1113, 3113, 132113, 1113122113, . . .

There is a rule behind the way this sequence is generated. The challenge is to keep asking new questions of the sequence, trying to look at it in different ways until eventually you hit on a perspective from which you can see what makes it tick. I'm going to let him sweat a little before I tell him the secret – because if he does get it, his brain will get that rush of adrenaline that I crave as I sit scribbling in my yellow pad all day.

I give him another sequence as well, just in case he gets the first one too quickly:

2, 3, 8, 13, 30, 39, . . .

My own work is dedicated to trying to understand the following list of numbers:

1, 2, 5, 15, 67, 504, 9,310, . . .

I do know what this sequence of numbers is describing. It is the number of symmetrical objects made from the symmetries of one triangle, two triangles, three triangles, four triangles, five triangles, six triangles then seven triangles. So the nth number in the sequence is the number of objects with precisely 3^n different symmetries. The point is that I have no idea how this sequence continues past the seventh number. It pushed the limits of a computer and two colleagues of mine to calculate that there are 9,310 shapes with 3^7 symmetries. I am trying to find some underlying pattern to the way these numbers are growing which might in turn unlock the secret of what these different shapes look like. Is there, for example, a formula that will generate these numbers as one adds more and more triangles?

I decide that the second sequence is a little unfair on my neighbour. Before he spends too much time trying to find a pattern, I put him out of his misery. If he'd managed to find a formula that turned up 49 as the next number in the sequence, I would have recommended that he buy a lottery ticket next weekend. Although they started off looking remarkably like some of the Fibonacci sequence, these were in fact last Saturday's winning National Lottery numbers. I'm planning to use them in a few weeks' time for a presentation I'm doing at a local school in Hackney about pattern searching. My neighbour laughs, although I can see he's a bit annoyed.

But my trick contains a warning. The human mind is desperate to find patterns. It is why we are so obsessed with symmetry. Pattern implies meaning. But sometimes things can be random and without patterns.

If my neighbour had actually managed to identify some structure behind the lottery sequence, something like the Fibonacci pattern, that would be another warning. There are always several different ways to make sense of any finite sequence of numbers which can then be used to generate the next numbers in the sequence (though the 'rules' may be extremely convoluted). I recently read a beautiful murder mystery set in my department – *The Oxford Murders*, by Guillermo Martínez. Maths and murder seem to go well together; perhaps mystery writers feel that the cold logic of the mathematician's mind is perfectly suited to cooking up the perfect undetectable murder. In this particular novel each murder is accompanied by the appearance of a mathematical symbol. The greatest logician in the department takes up the challenge of trying to crack the next symbol in the sequence before the next murder is committed.

But he has this sneaking worry that there might be several different solutions. Which one will provide the next twist to the mystery? For example, look at the sequence 2, 4, 8, 16, . . . You will no doubt be

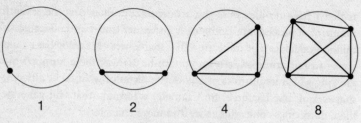

Fig. 41 Dividing circles.

convinced that the next number in the sequence is 32. But there is an equally compelling reason why 31 should come next. Take a circle, place two points anywhere on the circle, and join the points, dividing the circle into two (Figure 41). Now add another point, and draw lines joining it to the two points already there; the circle is now divided into four regions. Add another point, and connecting lines, and you find there are now eight regions. A fifth point and more lines take the number of regions to 16. But then, very unexpectedly, if you add a sixth point and a further set of lines, you can only get a maximum of 31 regions.

The formula for getting the number of regions with n dots looks at first sight like it should be simply 2^{n-1}, but mathematical analysis reveals that the right formula is in fact

$$\frac{1}{24}\left(n^4 - 6n^3 + 23n^2 - 18n + 24\right)$$

This example is a great warning to me. I have to find the right way to extend my sequence so that it really does describe the number of symmetrical objects as the number of triangles is increased.

I wish I could say that I made inroads into this problem by the power of logical reasoning. But the truth is that I stumbled on how to attack it purely by chance (an important factor in many break-throughs). I was giving a talk in Cambridge in the same department where Conway and Norton had shown me their Atlas containing all the building blocks of symmetry. After the talk someone asked me, 'Does that tell you anything about Higman's PORC Conjecture?'

I hadn't got a clue what this conjecture was. I nodded, trying to look knowing, and said it possibly could but I'd have to think about it. Afterwards I dived down to the library. In the 1950s, before all these wonderful objects like the Monster were discovered, the Oxford mathematician Graham Higman started to try to count the symmetrical objects that can be built from the symmetries of triangles or pentagons or other prime-sided polygons. In the periodic table of symmetry, the group of rotations of a prime-sided polygon is one of the simplest of the elements. The question Higman started to investigate was how many molecules could you build from copies of one of these atoms of symmetry? Higman knew that if he wanted to know how many objects there were with $p \times p \times p \times p \times p = p^5$ symmetries, where p is a prime number, then a simple set of formulae would tell him. He simply had to plug the prime number into one of the formulae and out would pop the number of objects you can build with p^5 symmetries. The choice of formula depends on what the remainder is when you divide the prime by 12. For example, for all the primes which leave remainder 5 on division by 12, i.e. 5, 17, 29, 41, 53, . . . , the formula is $2p + 67$. So if you want to know how many objects there are with 53^5 symmetries, you feed $p = 53$ into the formula, which gives the answer as 173.

The question then is whether, as we increase the number of copies of the prime building block we are using, there will always be a nice equation that will tell us the number of symmetrical objects that can be built from this atomic symmetry. Higman's PORC Conjecture claims that there will always be a polynomial expression that will tell you the answer. A polynomial expression in the prime p is something like $4p^3 + 17p^2 + 7p + 5$, where you take a combination of powers of p. For example, the number of objects with p^6 symmetries is given by a set of quadratic polynomials, equations involving taking squares of p. For the prime $p = 53$, the formula for the number of possible symmetrical objects with 53^6 symmetries is got by feeding $p = 53$ into the formula $3p^2 + 39p + 414$. Mike Vaughan-Lee in Oxford and Eamonn O'Brien in Auckland have discovered a more complicated quintic polynomial, one involving fifth powers of p, for counting the objects with p^7 symmetries. But are there such formulae to count the number of objects with p^8 symmetries, p^9 symmetries, p^{100} symmetries? We just don't know.

What I discovered the day I gave that seminar in Cambridge is that I'd been developing tools which might be the right ones to answer this question. As a way into the world of symmetry, I'd been exploring

things called zeta functions. The German mathematician Bernhard
Riemann first introduced this object into mathematics in 1859 as a
powerful way to try to find order in the chaos of prime numbers.
When you look at the sequence of prime numbers there doesn't seem
to be any simple rule or pattern to help you to predict where the next
prime will be: 2, 3, 5, 7, 11, 13, 17, 19, 23, . . . Riemann's zeta function
revealed a subtle structure underlying the primes which explains something of how the primes are laid out through the universe of numbers.
I've been trying to see whether my zeta functions can also be used to
understand any patterns in the wild world of symmetry.

For both primes and symmetries, zeta functions act as black boxes.
They are built from a formula which binds together the numbers you
are trying to understand. The hope is that the zeta function will reveal
new insights into the numbers of symmetries. It provides a way of
getting from part of the mathematical world where chaos seems to
reign to a completely different region where one can start to pick out
patterns. The meeting I'm heading to in Japan is bringing together a
small band of mathematicians who share the same obsession with zeta
functions.

I'm amazed that after all this, the passenger sitting next to me is still
looking quite interested. I look down to check the reading material
he's brought with him, in case it's yellow and he turns out to be
another of the mathematicians going to the conference. It turns out
he's clinging to a copy of the Bible. I feel I can't very well not ask him
what he's up to after he's sat through the details of my research. With
some trepidation, I put the question. 'I'm off to a missionary in Japan.
Do you know what Intelligent Design is?' My heart sinks as he tries to
convince me that the 'logic of math is just proof of the existence of
God'. The man sitting on the other side of me looks desperate to avoid
small talk with either of the other passengers in his row: the religious
nutter and the mad mathematical missionary. Our arrival at Narita
International Airport can't come soon enough.

As we come in to land, my neighbour eventually admits defeat with
my sequence of numbers and asks to be put out of his misery: '13, 1
113, 3 113, 132 113, 1 113 122 113, . . . so what comes next?' Conway,
the ultimate pattern searcher, was once set the same challenge and it
kept him baffled for months, but he wouldn't let anyone tell him the

secret. This is the sort of sequence that young kids actually find easy to 'get' because their view of the world is uncluttered by the complex patterns that adults are looking for.

So here's the answer: each number describes the previous number in the sequence. The first number consists of one 1 and one 3, which is written as 1113. This number can be described as three 1's and one 3, which becomes 3113. So the next number is 311311222113. Tomer saw it quite quickly when I first asked him. 'It's like a poem we did in school,' he said, and challenged me back to read the following out aloud:

> 11 was a racehorse
> 12 was 12
> 1111 race
> 12112

As we collect our luggage I can't resist teasing my missionary neighbour with another puzzle: 'If you like a challenge, prove that you'll never see a 4 appear in that sequence.'

Green trousers and green tea

My Japanese host is the mathematician Nobushige Kurokawa. I've read many of his papers and feel I know him already, but we have never met before. I haven't got a clue what he looks like, so when I emerge from customs at Tokyo I'm looking out for a board with my name on it.

I can't see my name anywhere. After a while I start to get a little concerned. I've come with no phone numbers, no contact details, just Kurokawa's email address. This is not a good start. But then I spot it – there's a man wearing a bucket hat walking around holding a board with a ζ painted on it. It's the Greek letter zeta. He's looking equally concerned, trying to pick out an unknown mathematician from all the Westerners coming through arrivals.

'Ahhhh yes, Professor du Sautoy. I should have spotted you. I carry a zeta and you are dressed in green.' I look a bit puzzled. I am wearing green trousers and a green hoody top. 'Green is the colour of zeta. It is the photosynthesis of mathematics, taking in light and giving out life.'

We hit it off immediately. His English is shaky and my Japanese non-existent. But our shared mathematical bond gives us the feeling of an ancient connection. Professor Kurokawa has a wonderfully eclectic perspective on mathematics and its relation with the outside world. In addition to being a great mathematician, he also seems to be something of a mathematical mystic – a Japanese Pythagoras.

'du Sautoy-san. You come in an auspicious year for zeta. It is 146 years since Riemann discovered his Hypothesis about zeta.' 146 sounds a bit arbitrary, I say. 'Not at all. 146 is twice 73. 73 in Japanese character is *nami*, which means "wave".' Like tsunami, meaning big wave. 'Zeta gives us waves to explain the primes. So 73 is zeta's number. 73 years ago Siegel made his great discovery of the formula for calculating the zeta function. So maybe we hit another peak this year for the story of zeta.'

Pure numerological nonsense, but wonderful. It is this playfulness that makes for a great mathematician. I recognized it in the way Conway does his mathematics when I first saw him in action on my visit to Cambridge. It is a rare trait, though: mathematicians often take refuge in the formal character of the subject and won't risk any levity.

The conference venue is on Okinawa, an island at the southern tip of Japan. Before we fly south we stop for something to eat. 'Mathematicians divide into two camps: those who love sweets and those who hate sweets. du Sautoy-san, which camp are you?' I can see by his chubby physique that we are in the same camp, so at 10 in the morning we go and feast on a strange assortment of green tea sweets, the perfect antidote to jet-lag.

The flight to Okinawa is exciting for three reasons. First, I get to sit next to Kurokawa and talk about zeta functions for three hours. Second, we have a fantastic view of Mount Fuji out the window. And third, we are on the Pokemon Plane. The inside and outside of the plane is festooned with an assortment of Pokemon characters that would send my son crazy. I am a little disappointed that the air hostess isn't dressed as Pikachu.

I'm particularly keen to talk with Kurokawa about a problem I've been battling with, related to my zeta functions. For some years I've been working on a conjecture about certain types of zeta function. I've broken it down into pieces. There are six different types of zeta function. For five of them I can prove my conjecture, but despite concerted efforts I can't find a way into the sixth. It has been bugging me for some time. It's like navigating an island. If I head north, I can see a river I can sail along to get to the coast. To the east the terrain is easy going. I can see a huge mountain to the west, something called the Riemann Hypothesis, one of our greatest unsolved problems. If I assume that the Riemann Hypothesis is true, I can make it past this mountain and I know that the terrain beyond is navigable. The trouble is that when I turn to the south, I hit impregnable jungle. All the tricks that helped me to navigate my way in the other directions are no use in this jungle.

I've been writing a paper on the problem, but it's stalled. Maybe I should send it off as it is: it would still be a good contribution to the literature on the conjecture. But it would be incomplete. One of the traits of a mathematician is an addiction to perfect, complete solutions. Riemann left many things unpublished because he felt that they were unfinished, and many of them went up in smoke when his housekeeper cleared out his office after he died suddenly at the age of 39. The formula Siegel discovered 73 years ago, mentioned by my host, was rescued and pieced together from Riemann's unpublished notes.

Kurokawa has read the preprint that I've prepared giving the results to date. He begins to explain how the zeta functions I am looking at fit into a framework that he considered some years ago. I read these papers but hadn't thought they were relevant. But during the three-hour flight he shows me why his language can be used to capture my zeta functions. I begin to go into a slight panic. Does this mean that Kurokawa's papers already prove what I've been working on for the last five years? I try to grasp the implications of what he is scribbling on the pages of my notebook. I looked through his papers for inspiration on how to tackle that sixth case. What did I miss?

By the time we touch down on Okinawa I realize that there is some overlap in the work I've done, but it isn't completely subsumed by Kurokawa's papers. Indeed, he can see that the case I'm left with is genuinely challenging. On the downside, he hasn't a clue as to how to tackle it.

Okinawa is a holiday island where young Japanese come to the beach to sunbathe and scuba dive. But like a team of miners descending into the darkness, we shall be in a lecture theatre from dawn till dusk, cut off from such blissful surroundings.

The gathering is small. There is a Russian, an Israeli, a German, an American, fifteen Japanese, and me. It feels like a gathering of some far-flung tribe to share stories of lonely wanderings across the mathematical globe in search of new lands. The dress is informal, veering on the dishevelled. Although we all speak different languages culturally, mathematically we are all on the same wavelength. Everyone has come to present their use of the language of zeta functions to reveal patterns and structure in different areas of mathematics. The talks are all in English. Despite the universal nature of the mathematical language, the words that frame the mathematics are important in bringing it alive.

Increasingly, the medium of choice for presenting the maths is the overhead projector although it is often abused in attempts to impress the audience with lots of results which just flash before your eyes in an array of different coloured inks. I'm as guilty of this as the rest. There is something about the pace of chalking up theorems and equations on a blackboard that is more in tune with the speed at which I can assimilate the ideas. On the downside, the lecturer does spend most of the talk with his or her back to the audience, so on balance I prefer the overhead projector. On the second morning of the conference, it's my turn to explain how I've been using zeta functions to count symmetrical objects.

Black box

The first zeta function was investigated in the middle of the nineteenth century by Riemann, who was interested in mixing the zeta function with new numbers called imaginary numbers. Like some alchemist, his belief was that the strong cocktail of ingredients would create some powerful mathematics. What emerged from his mathematical cauldron was a new way to understand prime numbers.

Riemann had read about these numbers as a child in the school library where he used to hide away from his classmates, terrified of most forms of social interaction. The security of mathematics was like the cupboard under the stairs where he could hide, protected from the pressures of the outside world. He had understood from a very early age that the primes represented one of the deepest challenges to the pattern searcher. The list of primes doesn't seem to possess any logic or order that might help you chart a course through them. As they continue on their way to infinity, they look no more ordered than the sequence of lottery ticket numbers with which I teased my fellow passenger. Guessing where along the number line the next prime will fall has baffled generations of mathematicians since the Ancient Greeks first studied them.

Riemann discovered in his thirties that the zeta function gave him a powerful new way of looking at the primes. It acted like a bilingual dictionary, enabling properties of numbers to be translated into geometry. Ever since Riemann's discovery, variations on the zeta function have been exploited as a way of revealing patterns and structure in mathematical settings where at first there just seemed to be mess and disorder.

The zeta function is something like a black box. Even when mathematicians know all the details of how to construct this black box, they are still left with a sense of wonder that it can reveal so much. Its construction depends on binding the infinite number of prime numbers together so that one is looking instead at a single object. It's like finding a way to analyse the overall structure of a musical symphony rather than studying it note by note. Riemann's extraordinary revelation is that the formula for the zeta function binds these mysterious numbers together in such a way that it is possible through analysing the formula to glean something of the secrets of these numbers.

Until a few years ago, no one had considered the power of the zeta function to reveal anything interesting about the world of symmetry. I was lucky that my apprenticeship as a PhD student coincided with the discovery that staring at symmetry through the eyes of the zeta function helped you to see things that no one had seen before. This new perspective was made possible by Dan Segal in Oxford, who would become the supervisor for my doctorate, and Fritz Grunewald in Germany.

What I've discovered is that zeta functions can be used to reveal certain patterns in the numbers I'm trying to understand. For example, what is the secret behind the sequence of numbers that starts 1, 2, 5,

15, 67, 504, 9,310, . . . ? These numbers count the number of different symmetrical objects that have 3, 3^2, 3^3, 3^4, 3^5, 3^6, 3^7, . . . symmetries. My zeta functions reveal something about how this number sequence continues. I've discovered that the numbers in the sequence obey a rule rather similar to the rule for creating the Fibonacci numbers.

The rule that generates the Fibonacci numbers is very simple: any number in the sequence is the sum of the previous two numbers. So once you know the first two numbers, you're away. Using the zeta functions of symmetry, I have proved that the same thing applies to the numbers I'm interested in. My proof tells me that there is a simple rule that generates the next number in the sequence 1, 2, 5, 15, 67, 504, 9,310, . . . Mathematicians call sequences generated by such a rule recursive sequences. They have a sort of computer program which generates them.

The only trouble is that my analysis doesn't tell me *what* the formula is. For the Fibonacci sequence, knowing any pair of adjacent numbers is enough to generate the next number. Although I have proved, using these zeta functions, that my numbers obey a similar rule, the proof doesn't say what the rule is, or whether it is 10, 100 or 1,000 numbers I need to know to generate the next number in the sequence. My discovery might look rather useless – and in a sense it is, because I can't use it to discover the next number. But it means that at least there is a pattern there to discover, that the numbers are not completely random but depend on each other in a way that is similar to the way that the Fibonacci numbers are all interrelated. It's quite striking that you can get enough insight to know that such a formula must exist without actually constructing it. This is a characteristic of many bits of modern mathematics: one can analyse a setting to prove the existence of certain structures without actually being able to construct them explicitly. It's a bit like discovering DNA but not yet having the tools to sequence the DNA explicitly.

My proof has at least shown that the infinite sequence of numbers is captured by something finite. It is like the difference between π, which as a decimal number looks completely random, and $1/7$, whose decimal has a clear pattern, namely 0.142 857 142 857 . . . , where the same six numbers are repeated over and over. My discovery reveals that there are similar patterns at work, and my numbers aren't completely wild, like π.

Before I know it, it is 11 a.m. and I have to finish my talk. Unusually, there are lots of questions. This is the advantage of a small conference. Generating discussion at a big meeting can be hard. Although many of the talks here are quite far from my own research interest, for me the most valuable thing is the prospect of picking up ideas which might just be transferable to my project. I have used my zeta functions to help me count how many symmetrical objects can be built from the symmetries of more and more prime-sided shapes. But there are lots of things I don't yet understand about what these zeta functions are telling me. So I'm hoping that seeing how others have used their zeta functions to see new structures might help in my own quest.

There are a lot of young Japanese graduates who are presenting their work for the first time. It is a terrifying moment when you have to pull your head out of the journals you've been studying for the last three years and stand up in front of your peers and superiors to present your contribution. Some of the young students' work is well received, but a couple suffer utter humiliation. 'Any questions?' 'Just a comment . . . I think if you look up my *Acta* paper of 1994 you'll find that I've already solved this problem.' Three years' work down the drain – proving a theorem that's already been proved, every mathematician's worst nightmare.

Although the days are spent at the blackboard or sweating over a hot projector, the evenings provide a chance to wind down over some sake and exotic food. The locals take us to a small bar on the corner near our hotel. Fortunately, I have read in the guidebook the advice to bring socks with no holes. A visit to Sock Shop at Heathrow has spared me the embarrassment of revealing my big toes as we sit cross-legged at the table for dinner, our shoes left at the entrance to the restaurant. The evening ends with Kurokawa expounding more of his wonderful mystical theories.

Mathematical expeditions

The organizers of every maths conference like to include a short excursion to give the participants a break from the onslaught of equations. At larger conferences the logistics of such trips are quite awe-inspiring. At the last International Congress of Mathematicians that I attended, the whole of Beijing was brought to a standstill by the government while four thousand mathematicians were transferred from the conference centre on the outskirts of the city to a feast held at the People's Palace in Tiananmen Square. At a conference that I co-organized in Durham, we decided to bus all the mathematicians to one end of Hadrian's Wall and pick them up in the late afternoon after an eight-mile walk along it. We got some strange looks from other visitors in the car park as a hundred and fifty badly dressed individuals emerged from coaches, babbling strange sentences full of pro-*p* groups and Lie algebras. All the conference participants had to do was to follow the wall until they met up with the buses, but we still managed to lose a few of them along the way.

I spent the whole 24 hours of my birthday one year travelling between conferences in Russia on the Trans-Siberian Railway accompanied by a hundred Russian mathematicians. As we entered the carriage, one of them was determined to sit next to me. He produced a book. 'I never meet before native English speaker and your assistance is much needed with six problems.' I assumed it was a maths book he was putting in my lap, but the title of the old and tattered tome was *1000 Jokes*. 'I've understood 994 of them but my English is not good enough to comprehend the last six.' There were six tiny bits of paper marking the pages in question. No wonder he was having trouble – the book was ancient and the jokes highly obscure. To his great disappointment, I managed to sort out just one of the six, and that required reading the joke out loud in an extremely posh English accent so that you could hear an obscure pun on words. Mathematics, it seemed, was better than archaic English humour at crossing cultural divides.

Recently I attended a conference in Assam. The mathematics was punctuated with a weekend trip to a rhino reserve. The rhinos in the early morning mist were a stunning sight, but not for me the most lasting memory of that trip. The limited accommodation in the lodge next to the reserve obliged us to share double beds. The slight anxiety of being paired up to sleep under the mosquito nets with Dan, my ex-supervisor, got translated in my dreams into a huge black dog clambering into the bed. I awoke to find myself in some Oedipal act physically assaulting Dan as he desperately tried to calm me down.

This afternoon, the Japanese mathematicians have laid on a trip to visit a distillery where they make Awamori, Okinawa's version of sake. Kurokawa is apparently terrified of the water, so the alternative outing, a mathematical scuba-diving expedition, was vetoed. At the end of the visit the owner of the distillery proudly presented us with miniatures of the beverage as a memento of our visit. When I pointed out that it was a shame it was only 30 per cent proof, and not a prime number, the owner suddenly whisked all the bottles away. I was a little nervous I'd offended our host, and I was getting some rather angry looks from the local visitors, when he suddenly reappeared with a whole new stash of bottles. '43 per cent,' he proudly announced, 'a prime number, I think!' Suddenly my fortunes were transformed, and I was heralded as 'prime-number-san' for the rest of the day.

This evening we have our conference dinner, another of the rituals of all such gatherings, when the tribe eats together. An Italian restaurant is an exotic end to the meeting for the Japanese participants, but after a week of eating sea-urchin salad, pig's ears in vinegar, bento boxes and sashimi, it seems odd to be eating pasta and drinking red wine.

During the dinner, one of the Japanese participants brings over one of the corks from the red wine we are drinking: 'bin 901' is printed along its side. Laying the cork before me, he says, 'Not a prime number!' I'm not sure whether he is expecting me to perform the same trick as at the Awamori distillery. 30 was obviously not prime, but this number is more difficult. I try a few small numbers, but given the confident air of the bearer of the cork, 901 is clearly going to be divisible by some larger primes. Soon someone at our table has got it – it's divisible by 17. Even that I find difficult to do in my head.

I think two different types of mathematical mind are illustrated by the incident with the cork. There are those who look at a number and immediately start trying to work out whether it is prime. Despite my love of prime numbers, I have never felt the need to do this. The other type of mind will look more for underlying structures and connections. Both are useful skills. The ability to crack a great unsolved conjecture quite often goes with the first. But the ability to come up with the conjecture in the first place, to have a new vision about how things might look, goes with the second.

I'm curious to know whether my hosts believe that there is a difference between the mathematics produced here in Japan and research in the West. Do the pictorial scripts of China and Japan create a different dialogue between mathematics and language? If your native script focuses more on pictures, does it affect the way you express mathematics? The numerological games that Kurokawa plays with language reflect a different way of looking at the world. My experience of working in Israel has been that the fantastic mathematical heritage of the Jews owes something to the Talmudic art of making strange connections between different sections of the Torah.

Some suggest that the extraordinary work ethic of the Japanese, which we witnessed during the conference, actually stifles the discovery of underlying structure. The lazy mathematician who is forced to find a short cut can often unearth an internal logic that is missed by someone with the perseverance to batter their way through endless calculations. But this seems rather stereotypical of Japanese culture, although it was the Japanese themselves who offered it up.

Despite our huge cultural differences, I think that Kurokawa and I have a very similar outlook on the mathematical world that is independent of our national heritage. And it is one of the things that draws me towards mathematics. As the famous mathematician David Hilbert once said, 'Mathematics knows no races . . . for mathematics, the whole cultural world is a single country.'

Mathematics and kabuki: theatres of the elite

Back in Tokyo, Professor Kurokawa takes me to a kabuki theatre. It is an extraordinary experience. The stylized, formal nature of the performance gives it a magic that a naturalistic drama can never have. It has its own inner logic and rules which the actors and even the audience adhere to. As each actor appears on stage, the knowledgeable members of the audience cry out the actor's stage name or even their kabuki number, the number they are assigned when they enter the profession, like a player in a football team. Shout out in the wrong place, and the ritual is destroyed.

There are intriguing resonances here with the world of mathematics. The kabuki actors have accepted the formal boundaries of their world yet are still able to be highly creative. As a creative process, doing mathematics can often feel like a theatre improvisation. You set up a tableau with conditions for collisions of ideas and then let the thing run. Very often it goes nowhere, but sometimes there is a dynamic created that clicks. Like the rules of a theatre game, the conditions push you in extraordinary, unexpected directions that too much freedom would stifle.

When the producer and director Peter Brook talks about his work in the theatre, he could easily be discussing the life of the mathematician: 'Small means, intense work, rigorous discipline, absolute precision. Also, almost as a condition, they are theatres of the elite.' Brook's last sentence highlights one of the other similarities between experimental theatre and mathematics: they both play to small audiences.

Tonight is rather special for the kabuki theatre, for they are introducing the audience to a new member of the troupe, the six-year-old son of one of the actors. The pride in their new member is evident. From now on he will be called White Hawk. I realize that our conference in a way has also been a celebration of the younger PhDs, who are the new blood we depend upon. They are our children, the people who will keep our ancient art alive and take it to new realms.

My trip to Japan ends with a journey to Nikko. The town boasts a stunning collection of Shinto shrines and Buddhist temples. The carvings and colour are stunning. But as I pass through one of the gateways into the courtyard of one of the shrines, I notice something rather odd. The gate is supported by eight columns decorated with a beautiful lattice of symmetrical patterns. All the columns are identical except for one, on which the pattern is upside down. It completely shatters the beautiful symmetry of the gate.

I ask Kurokawa about this. It is a deliberate decision, he says. It is a common feature of much of Japanese architecture, for the same reason that Arabic carpet makers deliberately weave a fault into their designs, for to achieve perfect symmetry is liable to anger God. The fourteenth-century Japanese *Essays in Idleness* articulate the ethos at the heart of the gate in Nikko: 'In everything . . . uniformity is undesirable. Leaving something incomplete makes it interesting, and gives one the feeling that there is room for growth . . . Even when building the Imperial Palace, they always leave one place unfinished.'

Perhaps this is how I should see my theorem on the zeta functions of symmetry. I have completed five cases, but the sixth remains incomplete. I had hoped that this trip to Japan would help me to complete the last piece in the jigsaw. But I shall take that column at the temple in Nikko as I sign that I should send my paper off as it is, and move on.

December: Connections

Luck favours the prepared mind.

LOUIS PASTEUR

5 December, Max Planck Institute, Bonn

The Max Planck Institute in Bonn is one of my favourite places. It's where I've made some of my most exciting mathematical break-throughs, the sort that give me the buzz and rush that I do mathematics for. One of these breakthroughs dramatically changed my view on how to count the number of groups it's possible to build out of prime-sided shapes.

I come to the Max Planck Institute several times a year for a week of white-heat brainstorming with my collaborator Fritz Grunewald, one of the creators of the zeta function as a tool for studying symmetry. Bonn is perhaps the most boring city in the world, which makes it a great place to work as there is nothing else to do – a blank sheet of a city. The Institute has recently moved to beautiful new accommodation on the upper floors of the old post office at the centre of the city. One of the advantages of all the bureaucrats shipping out from Bonn back to Berlin is that lots of nice buildings became vacant.

Several years ago I had one of those flashes sitting in the same office where I am now. I was trying to phone Shani in London but I couldn't get through. The phone was permanently engaged. It must have been about eight in the evening.

While I was waiting for Shani to finish chatting, I suddenly had an idea how to construct a new object whose group of symmetries could possibly illustrate completely new behaviour that we'd not seen before. I didn't physically build this object – that would be impossible, as it lives in nine-dimensional space. But I started to see how, using the language of group theory, I could write down rules for how all the different symmetries of this object interact with each other. I started scribbling equations on my yellow pad. First I needed to solve a little puzzle: essentially, finding a particular way of putting x's, y's and z's into a 3×3 grid – a mini sudoku with letters rather than numbers. Once I'd solved that puzzle, the thing looked rather beautiful and seemed to have just the right feel about it.

It was going to take some checking to make sure that the grid did what I needed it to. If I was right, this object would connect the world of symmetry with a completely unrelated area called elliptic curves. 'Flash of inspiration' well describes the feeling of revelation I had, because it really did feel like a rush of electrical activity coursing through my head. These flashes don't come often, but they are what most mathematicians live for. I can probably identify three such moments in my professional life to date. They are the equivalent of a footballer scoring the goal of the season. For the rest of the time, the process of discovery is more like running a marathon, where your continuous effort gradually accrues until the moment you pass the finishing line. Of course, even great goals depend on a solid build-up. This particular moment of discovery didn't come from nowhere, but emerged from months of groundwork laid here in Bonn.

I remember sitting in the office, feeling a little breathless. I wanted to tell someone what I'd done but there was no one around. I told Shani when I eventually got through to her, but she's not a mathematician and couldn't really grasp the enormity of what my result would mean to my research. She could tell from my voice that this was something exciting, but I needed someone who could share and appreciate the feeling I was having. I also needed to talk to someone who was capable of telling me whether I might be wrong.

I rang Fritz, one of the handful of people in the world who would understand and feel the excitement I was experiencing. We arranged to meet for a beer later that evening so that I could explain to him on paper the garbled words I was speaking down the telephone.

Fritz has a rather bashful air about him. His white shaggy hair makes him look like a rather affectionate pet. He has a gentle, low voice which often flips to hysterical laughter when we realize that we've made a crazy mistake in our calculations. But behind the wheel of his Mercedes he's a wild beast. I have to cling to the passenger seat in my attempt to concentrate on the maths we're discussing as Fritz screams up the fast lane of the highway.

We met for first time at a maths conference in the depths of the Black Forest, held at a mathematical retreat called Oberwolfach. Meeting Fritz was like finding a lost family member of the mathematical nomadic tribe. We discovered that we were passionate about the same things. Ever since that conference I've been coming to Bonn for a week at a time, and like two musicians Fritz and I jam together in the offices of the Max Planck Institute. By the end of the week we have quite often created something new. The months after are spent writing up the creations we cooked in our crucible in Bonn.

Although we speak the same language, we bring very different things to the collaboration. Sometimes it's as though we each have half of the same ladder, each with some of the rungs. We can't do any climbing by ourselves, but once Fritz and I put our two half-ladders together we can both get to the top. Often ideas feel almost preconscious and, left unarticulated, never quite crystallize. Trying to tell Fritz about a hunch I've got can sometimes be the important push that gives life to the idea. There is a lot of unspoken dialogue between us: grunts, hands waving in a desperate attempt to show Fritz the structure I'm seeing. Often I'll find that Fritz already has the language to articulate the structure I'm struggling to put into words.

It is a very fragile process. People often assume that we must all be doing mathematics by email and there is no need to meet. But our brand of collaboration could never be done electronically. For a start we often sit for hours, quietly thinking to ourselves, saying nothing, every now and again scribbling something down. But then a single word spoken can spark something in the mind of the other. Looking Fritz right in the eye, waving my hands and grunting is not something that can be replicated by email.

The whiteboard is the best canvas for collaborating, despite my aversion to it for private exploration. The yellow pad is a private space. On the whiteboard I can scribble things in red, blue or green, rub out silly things, try something new, draw a picture to express the thoughts beginning to take shape in my head. I sometimes regret the impermanence of the scribbles. When I get back home I can't always quite remember what we did. John Conway told me that one of his collaborators takes a digital photograph of the whiteboard before anyone rubs anything off.

There is a huge amount of trust involved in such a collaboration. It's delicate, really delicate. There are some people I have tried to collaborate with but found that they have too much of a competitive streak. You'll have been talking together, and then they come back after working all night saying, 'Look, look! I've cracked it!' It's understandable that people want to be first to solve a problem, to be top of the class, to get their name on a theorem. It's an important driving force in making progress, but it can be a great handicap in collaboration. A colleague told me that he shared an idea with others only when he was sure that it wasn't going to work. So I have to find somebody with whom I can have a really honest and open relationship if I'm to form a collaboration. It's a bit like a mathematical marriage. With Fritz I feel that we are on the same side. I can take risks with him and he doesn't judge me for the stupid things I often say. As in any meaningful and lasting relationship, trust is essential.

My ultimate goal is to count how many objects there are whose groups of symmetries are built from piecing together copies of the indivisible symmetry group consisting of the rotations of a prime-sided polygon. These are the objects with a prime power number of symmetries. As I learnt after the talk I gave some years ago in Cambridge, the PORC Conjecture put forward in the 1960s says there should be simple equations that give you the answer. This is certainly true if I count objects with p^5 symmetries. If I want to know how many objects there are with 17^5 symmetries, I simply plug the prime $p = 17$ into the formula $2p + 67$. But what if I count objects with 17^{10} symmetries? Will there still be a simple formula which, when you feed in $p = 17$, will tell me how many symmetrical objects there are with 17^{10} symmetries? After the discovery I've made here in Bonn, I'm not so sure.

The group of symmetries I cooked up when I was listening to the engaged tone while trying to contact Shani was a warning shot across the bows of my progress. If the objects with a prime power number of symmetries are put together in such a way that they satisfy an extra condition, which I'll call the elliptic condition, then the number of ways that you can build these special objects is not given by a simple formula, but depends on a completely different sort of problem from a branch of maths known as number theory rather than the mathematics of symmetry.

If you want to know how many objects there are with a prime power number of symmetries that also satisfy my elliptic condition, then you have first to solve the following problem: count how many pairs of numbers (x, y), where x and y are between 1 and p such that the polynomial $y^2 - x^3 - x$ is divisible by p. Finding solutions to these sorts of polynomials, called elliptic curves, is one of the most subtle questions of mathematics. The mysteries of these curves are at the heart of Andrew Wiles's solution to Fermat's Last Theorem. They are also at the heart of one of the seven so-called Millennium Problems, for each of which the Clay Mathematics Institute is offering a million dollars for a solution.

Solving this particular equation was one of the great achievements of the nineteenth-century German mathematician Carl Friedrich Gauss. In a diary entry, the young Gauss explains how to count the number of pairs of numbers (x, y) such that the polynomial $y^2 - x^3 - x$ is divisible by p. If p has remainder 3 when divided by 4 (the primes 3, 7, 11, ...) then there are always just p pairs. That looks nice and neat. But things are far wilder for those prime numbers p which have remainder 1 when divided by 4 (the primes 5, 13, 17, ...). For them there is no simple formula for the number of pairs.

Because of the connection to these strange objects in number theory called elliptic curves, I often call my group of symmetries 'my elliptic curve example'. This new group of symmetries led me off in a completely new direction. Suddenly, counting groups of symmetries became inextricably linked with counting solutions to complicated equations in number theory. It went against what everyone expected – which is why I like it so much. That's why I was so desperate to show Fritz my discovery.

When I showed Fritz my example that night, he also thought it smelt right. After our beer I went straight to bed. When I have a good idea in the evening I much prefer to go to sleep thinking I've made a breakthrough rather than staying up late into the night searching for any mistake. If there is one, it will still be there in the morning, and I can at least fantasize for a little longer about my 'breakthrough'.

Fritz, however, had stayed up late after our drink and convinced himself that what I had discovered really would do what I thought it would. Next morning he showed me a nice language I could use to analyse the symmetries. When I got back to London it took me months of careful checking and hard grafting to convince myself – and to be confident that everyone who would read the paper I was going to publish would be equally convinced – that the elliptic curve isn't just a mirage that disappears under closer scrutiny. There are lots of ways the elliptic curve could cancel itself out. But I was pretty sure it wouldn't.

But it is the memory of the birth in Bonn, rather than the hard work and nurture after the delivery, that still gives me that thrill. Before that night in Bonn, the group didn't exist. The next morning it was already one day old. I really do feel that it was an act of creation that brought it to life. There are an infinite number of other symmetrical objects I could have written down, but none would have been interesting. None would have had any special resonance. The role of the mathematician is to create something special from the huge palette of colours that mathematics offers. That is what makes mathematics an art.

And yet ... I can't help feeling that this group was sitting there waiting to be discovered in a way that a piece of music isn't. No one else could have created Bach's *Goldberg Variations*. Bach couldn't have been beaten to the composition by someone else. But the group I've discovered now looks a little like a new species of butterfly – it existed before it was discovered.

There is a huge amount of serendipity in the way we make mathematical breakthroughs. The mathematical world is hugely interconnected, so that answering one problem can give you an insight into another seemingly unrelated problem. Just as one can turn a cube and see a different face of the same object, a problem can be twisted and turned to reveal a new side to the question.

My night in Bonn had revealed a new facet of my subject. Turn the problem of counting the number of groups of symmetries, and from a new angle you are faced with the problem of trying to count solutions to strange equations called elliptic curves. The question Fritz and I are working on during this visit is to see what happens if we throw away this elliptic condition. Can I still get strange equations that I must solve to count how many groups of symmetries there will be in general?

This is not the first time in the history of mathematics that this connection has arisen. The notion of solving equations was the turning point in our understanding of symmetry. It would provide the language that mathematicians needed to be able to talk about what symmetry really is. However it would take slightly longer than one night for the connection to be understood.

Mathematical poetry: cracking the secrets of equations

Solving equations has an ancient heritage. Four thousand years ago, Babylonian mathematicians had started trying to solve quadratic equations that had arisen naturally during their attempts to calculate areas of land. One clay tablet from this era records a calculation of the perimeter of a field whose area is 60 square units and the length of the long side exceeds the short side by 7 units (Figure 42). This is the same as trying to find a number x which solves the equation $x^2 + 7x = 60$. Ancient Babylonian mathematicians found a way to reveal that $x = 5$ was the answer. Although they were without a mathematical language to formulate the question clearly, let alone articulate their method for finding the solution, the idea was there. It was the mathematicians of the medieval world who developed the Babylonian idea into a method which could be applied to all quadratic equations and is now taught to every schoolchild.

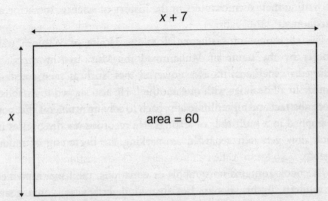

Fig. 42 Quadratic equations arise from calculating areas of land.

Following their conquest of the Persians in the seventh century AD, the new Muslim dynasty founded an empire that would become the hub of the world's cultural and educational development for the next half-millennium. While Europe stagnated, the cities of Kufah, Basrah and Baghdad were blossoming with libraries, museums, academies and mosques.

One academy in particular was to become the Mecca of intellectual life in the region, responsible for great advances in medicine, astronomy, philosophy and science. The House of Wisdom, the Max Planck Institute of its day, was founded by the caliph of Baghdad, al-Ma'mun. He wanted his city to become the new Alexandria and set about building a library and observatory. The first task for the scholars gathered at the new institute was to translate the huge number of ancient Greek, Latin and Hebrew texts that the Empire was amassing. Expeditions were being dispatched to gather up as many manuscripts as they could unearth that might have survived the destruction of the great library at Alexandria. The caliphs in Baghdad were even prepared to accept scholarly texts as part of peace treaties.

Although many of these texts suffered horribly in translation, the universal nature of the mathematical ideas meant that any errors that crept into translations of mathematical works were quickly picked up and corrected. The internal logic of the argument provided a self-correcting mechanism independent of the language the treatise was written in. As the knowledge of the ancient world was assimilated into the House of Wisdom, the scholars began to embark on writing their own chapters in the history of science, medicine and astronomy.

The champion of mathematics at the House of Wisdom was a scholar by the name of Muhammad ibn-Mūsā al-Khwārizmī. He believed in mathematics as a powerful tool 'such as men constantly require in all dealings with one another'. He also started to establish a more abstract and algorithmic approach to solving problems that could be applied in a multitude of settings. He recorded his discoveries in a book now generally regarded as marking the beginning of modern algebra.

The book contains no symbols or equations, the usual ingredients of modern algebra books, but instead describes in words general methods for solving equations. Although the methods are abstract, al-Khwārizmī does not lose sight of their power as a practical aid for his fellow citizens. Using a string of problems from legal disputes to the digging of canals, he justifies his belief that the ability to solve equations should be a fundamental skill for the ordinary man. Power, he advocated, lies with those equipped to speak the language of mathematics.

The mathematical name 'algebra' has its origins in the title of al-Khwārizmī's book *Hisāb al-jabr w'al-muqābala*. The Arabic word *al-jabr* was actually a medical term for the mending of fractured bones. As applied to mathematics, al-Khwārizmī wanted to convey the idea that an equation was masking numbers, and the algebra could restore or resurrect the hidden numbers, like a doctor mending a bone. For example, the unknown number might be called x (though in his book al-Khwārizmī describes all this in words – there are no symbols involved). An equation, however, gives you some information about that unknown x: for example, you might know that $x^2 + 2x = 3$. Al-Khwārizmī wanted to develop a method for manipulating equations so as to recover the hidden x.

So how is it possible to find any x's that will solve an equation such as $x^2 + 2x = 3$? Without the $2x$ term in there, we could solve $x^2 = 3$ immediately by taking the square root of each side. At first sight that extra term $2x$ makes things much more complicated. Al-Khwārizmī's strategy for solving these equations is first to make the equation look like the simpler one in which x only appears squared. So, by adding 1 to both sides and noticing that

$$x^2 + 2x + 1 = (x+1)^2$$

he shows how this equation can be written as the simpler equation

$$(x+1)^2 = 4$$

It's the same unknown quantity x, but now in a new equation. And al-Khwārizmī can solve this new equation because all he needs to do is take the square root of 4, namely 2, and subtract 1 from it. So $x = 1$ is the solution to this equation and to the original equation, $x^2 + 2x = 3$.

But there is another number that will solve this equation, and it is this second number that provides the first hint that there is some connection here with the mysteries of symmetry. This second answer was hidden from the scholars of the House of Wisdom because they hadn't yet discovered the power of a new sort of number: negative numbers.

It was mathematicians in India who put negative numbers on the mathematical map. Along with the concept of zero, they saw the potential of introducing new numbers to solve equations such as $x + 3 = 1$. They called these numbers 'debts' because they represented a useful way of denoting money that one person owed another. One of the first to write a treatise on the mathematics of these numbers was a seventh-century Indian mathematician by the name of Brahmagupta. As far as we know, he was the first to record that if you multiply a negative number by itself you get a number which is positive. Nowadays this is handed down to most schoolchildren as a piece of mathematical dogma, but Brahmagupta proved it by exploiting ideas similar to the algebra al-Khwārizmī would develop. Brahmagupta recognized that this discovery had implications for solving quadratic equations. It meant that every positive number has two square roots, one positive and the other negative. So $x = 2$ is a solution of $x^2 = 4$, but equally so is $x = -2$.

This is the sign that there is symmetry at work in these equations. The negative solution is a mirror of the positive one. Brahmagupta realized that more complicated quadratic equations will also have mirror solutions. For example, with the slightly more complicated equation $x^2 + 2x = 3$ considered above we find that the other mirror solution is $-2 - 1 = -3$. Brahmagupta was nonetheless still rather unsure of what this negative solution actually meant, given that these quadratic equations were helping to find the lengths of fences enclosing areas of land.

There is evidence that Brahmagupta was beginning to develop an abstract notation to articulate these equations. While the later algebra of al-Khwārizmī was still a book of words, Brahmagupta had started to experiment with using the initial letter of various colours to represent unknowns in equations. But this fledgling mathematical language did not blossom until it was reinvented in Europe around a thousand years after Brahmagupta's death. Even the concept of negative numbers was one that would not take hold in Europe for centuries, and without these numbers the mirror solutions would remain in the shadows. Negative numbers and zero are now so much part of our daily lives that it seems hard to believe that European culture took so long to accept these new numbers imported from the East. Negative numbers were associated with money lending, helping to represent debt. In medieval Europe, where usury was a sin, negative numbers were the embodiment of evil.

Although the Arabs may have been unaware of these mirror solutions, the newly acquired skills in manipulating quadratic equations had given the mathematicians of the House of Wisdom the confidence to see how far their new language might extend. Instead of falling back on the geometry and pictures of the Ancient Greeks, this algebraic language provided new means of gaining access to hidden solutions. They were developing a genuinely new type of mathematics that had the potential to yield much more than just solutions to quadratic equations.

If calculating the dimensions of fields used for farming gave rise to quadratic equations, then determining volumes of stone used for building gave rise to equations where the unknown quantities were cubed instead of squared. Was there some way to manipulate equations such as $x^3 + 2x^2 + 10x = 20$ to reveal the unknown x?

The eleventh-century Persian poet Umar al-Khayyāmī, better known in the West as Omar Khayyam, took up the challenge of cracking the cubic equation. But he was not working under ideal conditions. His home in the Persian town of Nishapur was under the control of the Seljuq Turks, who had invaded the region some decades earlier. While intellectual activity had been highly valued during the early years of the House of Wisdom, Khayyam found that he constantly had to compete against charlatans and astrologers for the attention of increasingly superstitious rulers. 'Most of our contemporaries are pseudo-scientists who mingle truth with falsehood,' he complained.

Khayyam was a real polymath. He wrote a treatise on music. He established one of the major observatories of the region, in Isfahan, from where he measured the length of the year to extraordinary accuracy, and his measurements led to a correction to the calendar in use at the time. He also wrote one of the classics of Persian literature, an epic poem of six hundred verses called the *Rubaiyat*. The title comes from the name of the poetic form that Khayyam uses. Each verse consists of four lines with the rhyming scheme AABA. Poets in this era revelled in the patterns and structure that could be woven through their poetry. Sometimes the third line of the verse is picked up to create the rhyming scheme for the next verse, BBCB. A cyclic symmetry starts to appear in the way the verses interconnect.

The rigid logic of its rhyming structure and its rhythmic patterns make classical poetry one of the literary forms that most resonates with the construction of mathematical proofs. So it is perhaps not surprising that Khayyam enjoyed the pleasures of mathematics as well as poetry. Although he made some progress with solving the cubic equation, a complete solution eluded him. 'Perhaps someone else who comes after us may find it out,' he wrote.

Equations involving cubes were as far as Khayyam was prepared to contemplate. The fact that they relate to the geometry of three-dimensional shapes gave him reassurance that there was some sense to his mathematics. For Khayyam, it was essential that there was geometry behind these equations: 'Whoever thinks that algebra is a trick in obtaining unknowns has thought it in vain. Algebras are geometric fact.' Contemplating equations with fourth powers, he dismissed them as meaningless because they would be describing geometric objects with more than three dimensions, and that was surely impossible. It would take a few more generations of mathematicians to sever the link between algebra and geometry and to see where that took mathematics.

Khayyam recognized that there were essentially 14 different sorts of cubic equation. He believed that a method that would work for an equation such as $x^3 + 2x = 5$ would also work if you took another cubic equation of the same overall pattern, where you just varied the numbers in the equation, for example $x^3 + 8x = 13$. Any equation of the form $x^3 + ax = b$ represented his first sort of cubic. An equation such as $x^3 + x^2 + 2x = 5$ was, in Khayyam's analysis, a different sort of cubic equation that might require different techniques. Because he did not know about negative numbers, he considered $x^3 + 2x + 5 = 0$ to be essentially different to $x^3 + 2x = 5$ because the 5 was on the other side of the equation. Once negative numbers were accepted as members of the mathematical menagerie, $x^3 + 2x + 5 = 0$ was recognized to be the same as $x^3 + 2x = -5$. Khayyam's ignorance of negative numbers meant that he ended up with 14 varieties of cubic equation. Eventually, once negative numbers and the number zero were added to the mathematical lexicon, these 14 cubics would be whittled down to a single generic type of cubic.

Although Khayyam made some progress, the complete solution to the cubic would not be found in the East. A century after his exploration of these equations, the great dynasty founded on the House of Wisdom came crashing down, when the Abbasid dynasty was brought to an end by the Mongols. It is thought that several million Muslims died in Baghdad while the major scientific institutions and libraries were destroyed. It was left to travelling European scholars and translators to pass the baton from the East to the fledgling academies of Europe.

Mathematical cock fighting

It was a colourful collection of Italian mathematicians in the sixteenth century who finally saw how to use the language of algebra to solve cubic equations. The discovery came at a time when European intellectual culture was swimming in the ancient achievements of the Greeks, now becoming known once again. These rediscovered works seemed to show that, after nearly two millennia, little mathematical progress had been made since the geometry of Euclid and Archimedes. Not only that, but excavations in Italy were revealing the ancient monuments of Rome. Renaissance Europe seemed unable to escape its ancient roots. When Italian scholars uncovered new mathematics that the Greeks and even the Arabs had never dreamt of, it came as a great fillip for modern European science.

The architect of this new mathematics, Niccolò Fontana, did not have a great start to life. He was almost killed at the age of 12 when the French invaded his home town of Brescia in 1512. During the slaughter of residents by Louis XII's troops, Niccolò was slashed across the face with a sabre and left for dead. He was rescued by his mother who tended to the horrific wounds her youngest had received. His jawbone was cut and his palate severed. The boy's wounds eventually healed, but he was left with a terrible speech impediment and from then on was always known by the nickname Tartaglia, 'the stammerer'.

In later life he grew a beard to try to mask the ugly scars left by the French invaders.

He may have been felled by the sword, but he would later be victorious in a battle of the minds. Shunned by schoolmates for his horrific appearance, Tartaglia turned to mathematics in order to escape the social pressures around him. Despite being self-taught, he found that he had a facility for the subject. He published a book explaining how mathematics could be used to predict the trajectories of artillery shells, and his work included the first tables for firing angles. But his great passion was for solving equations.

At the beginning of the sixteenth century it was generally believed that cubic equations were impossible to solve. This was the view of Luca Pacioli, who in 1494 had written what many regarded at the time as the definitive text on the state of knowledge of solving equations. Any breakthrough would have to come from outside the walls of academia, since Pacioli's view that the cubic was unsolvable was the received wisdom among most scholars. In 1534, battling away with these equations in the seclusion of his room, Tartaglia found the first chink in the cubic's armour. His secret was to exploit the idea of using cube roots as well as square roots. By using a combination of these different roots, he found that he could construct a formula to solve certain special types of cubic equation.

But Tartaglia discovered that he wasn't the only one who was claiming to have cracked the cubic. A young Italian, Antonio Fior, was boasting that he too possessed the formula for solving cubic equations. News spread about the breakthroughs made by the two mathematicians, and a competition was arranged to pit the two against each other. Bear baiting and cock fighting may have been the spectacles of choice for the peasant classes, but watching two mathematicians battle it out in mental combat was an entertainment more to the liking of the intellectual circles in northern Italy. Fior was extremely confident that he should be able to trounce the uneducated Tartaglia, convinced as he was that Tartaglia was just bluffing.

The trouble was that because European mathematics had yet to embrace the idea of negative numbers, there were many different types of cubic equation that needed to be analysed. As Omar Khayyam had already understood, there were 14 different sorts of cubic if one didn't use negative numbers. A method that worked for solving $x^3 = 5x + 1$

would have to be replaced with a new strategy when faced with $x^3 + 4x = 1$. Modern mathematicians armed with negative numbers would just rewrite $x^3 + 4x = 1$ as $x^3 = -4x + 1$ and solve it as they would $x^3 = 5x + 1$, with 5 replaced by -4. Without negative numbers, European mathematicians had to find an alternative way to crack these different sorts of cubic.

But the young Fior's method was not his own. The story goes that his teacher, Scipione del Ferro, had passed it on to him from his deathbed, in 1526. Del Ferro didn't want to take his secret with him to the grave, so he entrusted it to his student. But del Ferro had told his student how to solve only one of the 14 types of cubic equation.

On 20 February 1535 the mathematicians gathered at the great University of Bologna, then one of the largest and most famous centres of learning in Europe. Its reputation drew scholars from all over the region, just as the House of Wisdom had done centuries before in Baghdad. Public academic battles always drew great crowds, and the university was buzzing that day as Fior and Tartaglia arrived to lock mathematical horns.

Each contestant had been asked to provide 30 equations for his opponent to solve. The expectation was that 40 days would be needed for each to use his method to crack the 30 equations. A prize of dinner, paid for by the opponent, was offered for a solution to each equation solved. Although the problems that Fior had prepared for Tartaglia were cast in a variety of settings, from calculating the profit on the sale of sapphires to determining the height of a tree broken into pieces, all 30 of his equations actually reduced to the same type: the form $x^3 + bx = c$, where b and c took different numerical values in each problem. Fior, who had put all his eggs into one basket, was convinced that Tartaglia stood no chance.

Fior was almost successful in his unwitting strategy of basing all 30 questions on one sort of cubic equation. Although Tartaglia had made inroads into solving these equations, he had managed to find a way to crack only one of the 14 varieties of cubic, an equation that looked like $x^3 + bx^2 = c$, rather than those that Fior was planning to challenge him with. But spurred on by the forthcoming competition, in the early hours of 13 February 1535, just eight days before his duel with Fior was due to begin, Tartaglia managed to synthesize his ideas into a general method that would solve all cubic equations. By manipulating

the equations by making cunning substitutions, Tartaglia proved that
it was possible to change an equation of one variety into another. By
the end of his analysis he had discovered that there were really only
two different species of cubic that he needed to consider – and he
knew how to solve both.

Tartaglia managed to solve all 30 of Fior's challenges in a mere two
hours. In contrast to Fior's strategy of basing all his questions on one
sort of cubic, Tartaglia's problems for Fior ran through a whole variety
of different cubic equations. Fior was unable to extend his master's
method beyond the one sort of cubic that he had been shown how to
solve. Unable to see that the 14 varieties of cubic were actually examples
of two cubics in different guises, Fior was revealed for the mediocre
mathematician that he was. Despite his triumphant success, Tartaglia
declined the 30 meals he had won at Fior's expense.

News of Tartaglia's staggering victory spread quickly through the
corridors of the University of Bologna and beyond. One mathematician
was particularly keen to discover the secret to Tartaglia's success, and
began to press him to yield his magic formula.

The controversy of the cubic

Girolamo Cardano had a talent for getting into trouble. Tact was not
his strong point, and he was forever aggravating those in a position of
power. He had trained in medicine rather than mathematics, at the
University of Pavia. His desire for power led to his election as rector
of the university, a contest he won by a single vote. He was quite aware
of how unpleasant most people found his aggressive political style, but
he remained unapologetic:

> This I recognize as unique and outstanding amongst my faults – the
> habit, which I persist in, of preferring to say above all things what I
> know to be displeasing to the ears of my hearers. I am aware of this,
> yet I keep it up wilfully, in no way ignorant of how many enemies it
> makes for me.

Although a lawyer by profession, Cardano's father was a talented
mathematician, even advising Leonardo da Vinci on matters of

geometry. He died during Cardano's campaign for rector, but not before passing on his aptitude for mathematics to his son. He had taught Cardano the rigorous logic of mathematics in the hope that it would provide a great platform for a legal career. But the rebellious Cardano had other ideas. His mathematics had given him an understanding of the theory of probabilities, which he took with him to the gambling halls of Italy.

Cardano was one of the first to realize that there might actually be a way to predict the likelihood of certain numbers coming up when a pair of dice are thrown. He attempted to put into practice his analysis of the chances of rolling a double six, but an addiction to gambling soon took the place of rational analysis of the mathematics behind the dice, and Cardano ended up squandering the money his father had left him. One particularly desperate night he accused a fellow gambler of cheating him at cards. The maths might have meant that he stood a good chance of winning, but he couldn't admit defeat and instead drew a knife and slashed his opponent about the face.

None of this helped to secure the respect he needed to build up his medical practice. When the authorities discovered that he was illegitimate, it gave them the excuse to exclude him from the College of Physicians. Having pawned both his wife's jewellery and their furniture to fund his continuing gambling, Cardano was eventually forced into the poorhouse in 1535. Mathematics eventually came to his rescue. His talents had not gone unnoticed, and he was bought out of poverty and offered a position as lecturer in Milan. He continued to practise medicine, with some notable successes, but it was his writing on mathematics that began to secure his reputation.

Cardano was particularly interested in how to solve equations. Mathematicians believed that, unlike quadratic equations, equations involving cubes could not be solved using a magic formula. This was what Cardano had read in *Summa*, Pacioli's definitive book on arithmetic written in 1494. But then he heard the news that an unknown mathematician called Tartaglia had solved 30 cubic equations with amazing speed. Cardano knew that Tartaglia could only have done this with the aid of a formula.

Once you become aware that a solution is possible, that another mind has managed to conceive of a way through what had previously been thought impenetrable, the challenge is there to see if you too can crack it. Most mathematicians feel that if one person can work something out, then they should also be able to. After all, mathematical argument is such that it feels independent of the mind that created it. Once something has been proved, it starts to take on a concrete reality. But before that first breakthrough, there is always the nagging sense that there is no way through, that there is something essentially impossible about the task.

Mathematicians can't bear to admit defeat. The last thing they want is to have to be told the answer. So Cardano battled with the problem for several years, convinced that if there was a formula he should be able to discover it himself. By 1539 he could bear it no longer – he gave in. He sent this mysterious Tartaglia a request asking whether he could include the formula in a book he was writing about arithmetic. But Tartaglia was certainly not going to let anyone else publish his discovery, and he told Cardano that he was intending to publish his formula himself.

By now, Cardano was desperate to know the answer. He contacted Tartaglia again, promising not to communicate the formula to anyone if only Tartaglia would tell him. Again, Tartaglia refused. Cardano was incensed. What was the point of keeping the formula secret? Wasn't it Tartaglia's duty to share his discovery with his fellow mathematicians? He challenged Tartaglia to an open debate. But there seemed little point in this: after all, there was no dispute about whether Tartaglia had really made the breakthrough. Unlike some mathematical claims, the fact that he could solve the cubic equations he had been challenged with was proof that he had a formula. There was no onus on Tartaglia to prove himself further. He refused yet again.

Finally, Cardano saw that the way to tempt Tartaglia to reveal his secret was to offer him money. He wrote to Tartaglia, gently suggesting that a wealthy patron, the governor of Milan, was interested in sponsoring the great mathematician who had cracked the cubic. If Tartaglia was to come to Milan, Cardano said, he might be able to effect an introduction. Cardano's plan worked. Tartaglia was in desperate need of financial support. His meagre teaching position in Venice was barely keeping him in food and lodging, so he wrote to Cardano accepting his offer, and in March 1539 travelled from Venice to Milan.

According to Tartaglia's subsequent account of the meeting, Cardano was most hospitable, but kept pressing him to explain the secret of the cubic. Tartaglia, on the other hand, wanted to know when he could go and meet his rich new sponsor. Ever the schemer and manipulator, Cardano had planned Tartaglia's visit to coincide with the departure of the governor to the neighbouring city of Vigevano, some 50 kilometres outside Milan. 'We will have plenty of opportunity to talk and discuss our affairs until he returns.' Cardano began to press Tartaglia on why he had been so secretive about divulging his discovery of the formula for the cubic.

When you've made a mathematical breakthrough, there is always the possibility that this new idea might yield much more. Tartaglia could see that if the method he had devised could crack the cubic, perhaps it could also be extended to more complicated equations such as quartics and quintics – those containing terms in x^4 and x^5. He explained to Cardano that he didn't want to go public before he had at least followed up his belief, in case he was sitting on a mathematical gold mine. But for the foreseeable future he was fully occupied with teaching and with preparing a new translation of Euclid.

Cardano promised not to divulge Tartaglia's secret to anyone – he simply had to know for himself what the magic formula was. Tartaglia didn't believe him. Cardano was now going crazy: he was desperate to know the answer, having sweated for several years in unsuccessfully searching for the cubic's secret:

> I swear to you, by God's holy Gospels, and as a true man of honour, not only never to publish your discoveries, if you teach me them, but I also promise you . . . to note them down in code, so that after my death no one will be able to understand them.

That is Tartaglia's account of Cardano's promise, made three days after Tartaglia's arrival in Milan.

Tartaglia's patience was now at an end. He would ride to Vigevano himself to talk to the governor. But Cardano had named his price for the letter of introduction that Tartaglia would need. So under oath not to divulge the formula to anyone else and never to write it down for others to discover after his death, Tartaglia finally relented. 'To enable me to remember the method in any unforeseen circumstance, I have arranged it as a verse in rhyme,' he explained. The rhyme was rather long-winded and cryptic, but it held, he said, the key to his success in all the contests he had won. And so he began to pen the rhyme for the eager Cardano:

> When the cube and the thing together
> Are equal to some discrete number,
> Find two other numbers differing in this one ...

And so it continued in similar vein for 21 lines, explaining how to manipulate the equation until it gave up the secrets of its solutions. The poem concluded:

> These things I found, and not with sluggish steps,
> In the year one thousand five hundred, four and thirty.
> With foundations strong and sturdy
> In the city girded by the sea.

Cardano's relief was palpable: 'How well I will understand it, and I have almost understood it at the present.' But Cardano's relief as he pored over the poem Tartaglia had left him was matched by Tartaglia's own feeling of unease. Why had he just divulged his one great discovery, the formula that might open a way to formulae for all equations? He still didn't trust Cardano.

Instead of riding out to Vigevano, Tartaglia turned his horse towards Venice and headed home. But as he rode on, he became increasingly angry. He began to see how, with the lure of a rich patron, Cardano had duped him into revealing his treasured formula. By the time he got back to Venice he was utterly convinced that it was only a matter of time before Cardano broke his promise and published his discovery. When two new books by Cardano were published a year later, Tartaglia was certain that his darkest fears were about to be realized. But when he read the books he could find nothing about his solution to the cubic.

Cardano, despite his rather obnoxious character, had been true to his word and kept Tartaglia's formula secret. Well, almost. He couldn't resist discussing it with his best student, Lodovico Ferrari. Ferrari had originally been Cardano's servant, but when Cardano discovered that the 14-year-old boy could read, he promoted him to be his personal secretary. As time went by, Ferrari soaked up the ideas that Cardano shared with him. Now that the pair spent so much time together, it was only natural that Cardano would discuss Tartaglia's poem with Ferrari.

As he worked his way through the cryptic poem, Cardano began to grasp the method. The trouble was that when he then tried to apply it to certain equations there was a rather worrying glitch. In the middle of Tartaglia's poem was an instruction that sometimes resulted in you having to calculate the square root of a negative number. Cardano didn't know any numbers whose square was negative. What did this mean? The ancient formula for quadratic equations sometimes yielded square roots of negative numbers, but when this happened you just resorted to saying the equation was not solvable. But there was something rather strange about Tartaglia's method for solving the cubic. If you just ignored the fact that you didn't know what the square root of a negative number was, eventually, by the end of the poem, these mysterious square roots had disappeared, having somehow cancelled each other out, to leave a perfectly ordinary number which would solve the equation. Was there magic at work here? Had Cardano understood it properly?

On 4 August 1539, Cardano wrote to Tartaglia about the strange problem he was having. It's not clear whether Tartaglia was any clearer about the mechanics of what was going on, but he saw this as an opportunity to throw Cardano off the scent: 'I say in reply that you have not mastered the true way of solving problems of this kind, and indeed I would say that your methods are totally false.'

However, Tartaglia's worst fears about divulging his formula were about to be realized. The young Ferrari, barely 18, had discovered how to use the solution for the cubic to obtain a formula for solving equations that contained terms in x^4: quartic equations. Cardano was so impressed with the young man's discovery that he resigned his own position at the Piatti Foundation in Milan to make way for the young prodigy. But now he faced a dilemma. Because the method of solving quartic equations was based on Tartaglia's solution of the cubic, there was no way they could go public with Ferrari's discovery. Although Cardano had already broken part of his promise by discussing Tartaglia's proof with Ferrari, he still felt honour-bound not to publish anything. But that would mean denying his student the rightful praise he so deserved.

Cardano suggested travelling to Bologna to ask his colleague Annibale della Nave for advice about their dilemma. Della Nave proved a rather fortuitous choice. In his possession was a battered old notebook which had belonged to his father-in-law, Scipione del Ferro, the mathematician who had first cracked the cubic and had, on his deathbed, told his student Fior of his discovery. As Cardano and Ferrari leafed through the notebook they recognized the formula that Tartaglia had discovered, independently, some years later. Here was their way out: Cardano could now legitimately publish del Ferro's formula without any feeling that he was breaking his promise to Tartaglia. In the great opus in which he finally broadcast the solution for the cubic and quartic to the world, known as the *Ars Magna*, Cardano gives due credit to Tartaglia for his independent discovery. But it is del Ferro's contribution that gets the greatest praise: 'this art surpasses all human subtlety and the perspicuity of mortal talent and is a truly celestial gift'.

Unsurprisingly, the credit that Cardano gave him was not sufficient recompense for Tartaglia. His world was now crumbling around him. He'd missed out on writing about his own discovery of the formula, and now he'd been beaten to the solution of the quartic by an 18-year-old upstart. It was all too much.

In a vain attempt to rescue his reputation, Tartaglia wrote an account of his side of the story which included a string of venomous attacks on Cardano. Having stepped on many toes on his way to the top, Cardano was quite used to insults being thrown his way, but his young student Ferrari felt compelled to defend his teacher's honour. He wrote to Tartaglia, taunting him about his mathematical inadequacies and challenging him to an open debate. Such a contest with a relatively unknown mathematician would do nothing for Tartaglia's reputation. Beating Cardano in open mathematical combat would be a prize worth competing for. So Tartaglia wrote back to Ferrari trying to draw Cardano into the fray.

Letters flew back and forth between Venice and Milan. They were made public as the two men each attempted to win over the wider mathematical community. Insults were mixed with mathematical problems as Ferrari and Tartaglia tried to outdo each other. The range and variety of Ferrari's challenges to Tartaglia reveal that his solution to the quartic was no flash in the pan, but that he was maturing into a deep and philosophical thinker. In addition to the cubic equations, he challenged Tartaglia to solve problems in geometry 'proving everything', to illuminate passages of Plato, even to debate the philosophical problem of whether 'unity is a number'.

Despite his disdain for Ferrari, Tartaglia couldn't resist engaging with the 30 or so problems that he sent him, and he was gradually drawn into an extended correspondence. Philosophical debates about Plato and the concept of number were dismissed by Tartaglia as questions unworthy of a mathematician – not an uncommon belief among many modern mathematicians who are scornful of those who dedicate their time to the philosophy of the subject. In response to Ferrari's mathematical challenges, Tartaglia often claimed that his rival was angling for answers to questions he didn't know how to solve in the hope of stealing yet more of Tartaglia's ideas: 'It is a very shameful thing, to put forward such a question in public and not to know how to solve it.'

Ferrari responded by pointing to Tartaglia's lack of proof in presenting his solutions: 'Just like a forger you omit the part that matters, namely these two words "proving everything".' And Tartaglia's reluctance to debate deeper issues of mathematical interest, he said, reflects someone

who spends the whole time on roots, fifth powers, cubes and other trifles. If it were up to me to reward you, taking example from the custom of Alexander, I would load you up so much with roots and radishes, that you would never eat anything else in your life.

Things came to a head when Tartaglia was offered a prestigious position at the university of Brescia, his hometown. The offer was conditional, however, on his success in open debate with another mathematician chosen by the faculty. His heart must have sunk when he got a letter informing him that he was to travel to Milan to compete against Ferrari. If he wanted to get the lucrative job he had spent years trying to secure, he would have to swallow his pride and compete against Cardano's pupil. In any case, he believed that Ferrari was not truly equipped with the mathematical skills to put up much resistance.

On the morning of 10 August 1548, the two mathematicians came together in the beautiful gardens of one of the Franciscan monasteries in Milan. The open letters that had been exchanged over the previous two years had attracted much interest in the contest, and the garden was packed with onlookers, including a host of Milanese celebrities keen to witness mathematical blood being spilt. For Tartaglia these contests were his bread and butter, and he was confident that he would see off Ferrari's challenge. Tartaglia had only his brother there supporting him, but Ferrari was cheered on by a crowd of friends he'd invited.

As they locked horns, Tartaglia began to see that Ferrari had not been bragging when he said he knew the answers to all the questions he had sent Tartaglia. His young adversary, it turned out, had a far greater control of the formulae that would solve cubic and quartic equations. Tartaglia resorted to the poem he had concocted to help him to remember his method, but Ferrari was just too quick for him. Tartaglia resorted to firing petty shots across Ferarri's bow, criticizing his methods, in an attempt to knock Ferrari off his stride. As the day drew on, Tartaglia could see that it was a lost cause. Ferrari landed a succession of ever more telling blows, revealing how shallow Tartaglia's grasp of solving equations was compared with the young student's ability to twist and turn the formulae to his advantage.

When the crowd reconvened for the second day of combat, they learned that Tartaglia had fled back to Venice, preferring not to suffer the complete humiliation of a mathematical knockout. Ferrari was crowned victor and showered with offers of employment, including a request from the Emperor Charles V (whose grandfather had kicked out the Moors from the Alhambra) to tutor his son. Ferrari had his eye on making his fortune, preferring instead to become tax assessor to the governor of Milan. How many mathematicians have been lured away since then by the temptations of the city! Ferrari achieved fame with his formula for solving the quartic, made his fortune as a young man, but died aged only 43. He was allegedly poisoned by white arsenic at the hand of his own sister, who was after the huge inheritance she would receive on his death. She married two weeks after Ferrari's funeral only for her new husband to abscond with the money, leaving her destitute.

Tartaglia spent another year lecturing in Brescia, but after his igno-minious defeat the university decided first not to pay him, and later to terminate his position. He was incensed, but despite numerous lawsuits could get nothing out of the university for the work he had done for them. Crushed and penniless, Tartaglia finished his days in Venice.

Cardano was distracted from doing any more serious work by a series of disasters that beset his two sons. The eldest, Giambatista, had secretly married 'a worthless, shameless woman' who was interested only in extorting as much money as she could out of her now rich and famous father-in-law. Their relationship began to deteriorate, and the woman began openly mocking Cardano's son with claims that their three children were not his at all. Eventually Giambatista couldn't take the abuse any longer and poisoned his wife.

At his subsequent trial, the judge said that he would spare Giambatista from the gallows if his father would seek reconciliation with the murdered woman's family. Ever the money grabbers, the family insisted that forgiveness would come at a price that was well beyond anything Cardano could pay. Cardano never recovered from his failure to save his son from being first tortured in prison and then executed, on 13 April 1560.

To add insult to injury, Cardano's youngest son, who had inherited his father's passion for gambling, lost everything and resorted to stealing money and jewellery from his own father. Cardano reported his son to the authorities and had him banished from Bologna. In his autobiography, *De Vita Propria*, Cardano wrote that his four great tragedies were his marriage, the bitter death of his eldest son, the base character of his youngest son, and imprisonment. The last of these refers to the time he spent in jail at the end of his life at the hands of the Inquisition, accused of heresy. He had deliberately offended the Church, it seems, in an attempt to gain a place in history by writing a book praising Nero for tormenting the martyrs and for casting the horoscope of Jesus. It was not blasphemy, however, but his role in the story of solving equations that secured his fame after his death. Cardano committed suicide on 21 September 1576 – not through despair at the horrors he suffered in later life, but apparently to fulfil a prophecy he made some years earlier about the date on which he would die.

12 December, Max Planck Institute

Fritz and I have spent the week trying to understand how to find how many symmetrical objects can be built with a prime power number of symmetries. The lesson I have learnt from the breakthrough I made here in Bonn on a previous visit is that the number of symmetrical objects could depend on solving a completely different sort of problem than we first envisaged. The number could actually be the same as the number of ways of solving a set of equations. The question for us is this: what is the nature of the equations we will have to solve? Will they be as exotic as the wild elliptic curve that is at the heart of my breakthrough in Bonn? Or might they be simpler ones, like the equations Cardano and Tartaglia were solving?

Fritz and I have spent the week trying to analyse the sort of equations that might come out of counting all these groups of symmetries. The trouble is that the thing has started to explode into a vast problem that we can barely hope to master. I feel as though I have picked up a stone which has suddenly become the size of a mountain. Is it possible that we can hit upon some amazing idea that will embrace the huge complexity of the problem?

My head is hurting under the strain. I am exhausted at the end of each day. But instead of trying to master the mountain, we have now decided to go for something more manageable – one of its foothills. We've found a way to look at a small chunk of the problem, and we can even use a computer to do some experiments. I love it when this happens. I rarely get to use the computer in my work, but when I do it always feels to me like real science. There is work involved in setting up the experiment, but then the computer will mindlessly compute. And we shall get some answers! Rather than bashing our head against the whiteboard trying to come up with an abstract way through this world, we can set off on little exploratory forays to assess the lay of the land. The grand theory we are ultimately after will be like an understanding of the whole geography of a land, gained without visiting all of it. This first exercise on the computer, though, is like surveying what we can see around us. It will be important that the outcome is not a huge amount of random data with no discernible pattern.

It is a relief to take some time out to wait for the computer to produce some answers that might guide us. We head to the square outside to get a coffee while the computer sits calculating away. It is coming up to Christmas, and there is a strong smell of *gluwein* in the air coming from the Christmas markets across the city. My favourite time to come to Bonn is for *Karneval* in a few months' time, when the usually sober Germans don crazy costumes in a week of madness. I was here last year during *Karneval* week and saw Germans dressed as bananas and bumble bees walking round the streets of Bonn. No one looked as if they were particularly enjoying themselves, but everyone appeared seriously intent on getting very drunk. Fifty-year-old men stand at stalls eating sausages while dressed as a dog or a ladybird. Fritz claims that he doesn't get dressed up. That's a shame – I would love to have seen him in a banana suit.

When we go back upstairs to see how the computer has got on, we find that we've been asking it to do too much. It has run out of memory. Once again, we are on our own in the mathematical jungle. Whenever the going gets tough, I often try to recall those moments when complete befuddlement were suddenly transformed by the clarity of an idea. That particular night in Bonn is my touchstone. It helps me through the dark days.

From Bonn I decide to pay a visit to another of my favourite German cities, Göttingen. This Hansel and Gretel village was once the Mecca of nineteenth-century mathematics and home to two of my mathematical heroes, Gauss and Riemann. I spend the afternoon with a friend in the local cemetery. It sounded rather a macabre way to pass the time when he first suggested it, but it turns into a fascinating scientific pilgrimage.

Many of Göttingen's greatest scientists are buried in here. In addition to names and dates, the gravestones often bear the equation that made the name on the stone famous. It is on this trip that I decide what I want as my gravestone epitaph. I'm not saying that the discovery I made that night in Bonn is going to shake the world, but if I'm going to have anything on my gravestone, then it has to be the equation that defines the group of symmetries I concocted:

$$G = \left\langle x_1,x_2,x_3,x_4,x_5,x_6,y_1,y_2,y_3 : \begin{array}{ll} [x_1,x_4]=y_3, [x_1,x_5]=y_1, [x_1,x_6]=y_2 \\ [x_2,x_4]=y_1, [x_2,x_5]=y_3, \\ [x_3,x_4]=y_2, \qquad\qquad [x_3,x_6]=y_1 \end{array} \right\rangle$$

January: Impossibilities

Eliminate all other factors, and the one which remains must be the truth.

SHERLOCK HOLMES, 'The Sign of Four'

23 January, Oxford

I owe my marriage to Shani to knowing about palindromes. When I was a young post-doc I was looking for a flat in Jerusalem, but getting nowhere because I wasn't Jewish. For weeks I searched for somewhere to stay. I finally struck lucky when a woman poked her head round the door I'd just knocked on, looked me up and down and barked, 'So do you know what a palindrome is?' If you're not Jewish, then being an intellectual seemed to be the next best thing. I passed the initiation rite and got the spare room. My third flatmate became my wife.

Our first date together was something of a baptism of fire for both of us. Her friend had just had a baby boy. So Shani decided to take me to his *brit millah*, or circumcision. We sat around eating chicken drumsticks while the father and grandfather pinned the baby to the table and the *mohel* carved off his foreskin. To try to banish the frightening images from my mind I spent the remainder of our date explaining to Shani the rituals of my religion. Square ashtrays and round glasses were spun on the table top as I tried to initiate her into the secrets of symmetry.

Shani has gone back to Israel this week for another circumcision. I've decided to give this one a miss and escape to Oxford to bury myself in mathematics. Given my lack of progress in Bonn last month, I've decided to take a look at a completely different problem. I often have several problems on the boil at the same time. Sometimes a change of perspective helps when I return to a question after spending time on something else. On my first visit to Israel, 15 years ago, palindromes not only got me a flat but also started to make a rather intriguing appearance in the zeta functions I was studying.

Each group of symmetries has a zeta function attached to it. This zeta function can be described by a formula built from polynomial expressions. If the group of symmetries is not too complicated I can calculate the formula for the zeta function of that particular group. The curious thing that I noticed back in Israel is that every time I did one of these calculations, I always seemed to get polynomial expressions with a rather beautiful palindromic symmetry, for example

$$2x^6 + 4x^5 + 7x^4 + 3x^3 + 7x^2 + 4x + 2$$

The numbers in this expression form a symmetric pattern:

$$(2, 4, 7, 3, 7, 4, 2)$$

Just like words, formulae can form palindromes as well.

This sort of symmetrical pattern is highly unexpected. If you got a computer to generate random formulae, most of them wouldn't have this palindromic symmetry. Yet whenever someone took a new group of symmetries and managed to calculate its zeta function, the resulting formula had this palindromic symmetry, what mathematicians call a functional equation. But would this always be true? As time went on and more examples were added to the database, the evidence that it would got stronger and stronger. I couldn't see any reason why starting with an object full of symmetry should force the resulting zeta function to have this palindromic symmetry. It seemed right, but I was after a proof. How could I be sure that when I calculated the next example, this symmetry wouldn't disappear? Perhaps I was just looking at examples that had a very special structure. The trouble with mathematics is that evidence can often be very misleading. What looks like a very strong pattern can suddenly vanish before your eyes. That is why mathematicians are so obsessed with proof. For the other sciences, evidence is paramount, but mathematicians will put their trust only in proof.

Since my first visit to Jerusalem I have made many attempts to prove that this palindromic symmetry will always be there. The palindrome itself is not that important. For me this beautiful pattern is more of a beacon, hinting at a deep underlying structure which is itself responsible for the palindromic symmetry. My hope was that this beacon might guide me to a huge vista of structure that was manifesting itself through this palindrome. In recent years, though, I've come round to thinking that I might be seeing palindromic symmetry only because I chose the right examples to look at.

Suddenly, with the discovery of my new group of symmetries in Bonn, there was a chance that the palindromic symmetry would disappear. After all, it had completely changed my perspective on the PORC Conjecture. Perhaps it was malicious enough to destroy my Palindrome Conjecture. But I was in for a surprise.

A year after my discovery of this new group of symmetries, Christopher, one of my graduate students, sent me an excited email asking whether he could come up to London to show me something. My PhD students know that the best place to get my undivided attention is at my home. However, what Christopher had to show me that day would have captivated me in the most distracting of environments. He'd made a complete calculation of the zeta function of my elliptic curve symmetry group. My analysis of the group of symmetries had done enough to dig a tunnel connecting it with the world of elliptic curves. Now Christopher had built a permanent bridge.

But would this new zeta function have palindromic symmetry? As we looked at Christopher's calculation there seemed to be something wrong. The numbers didn't quite match up – almost but not quite. But then we spotted that something in Christopher's calculation that related to the elliptic curve might also contribute to the symmetry. We'd missed this bit. I rushed upstairs to get a book about these special curves, and soon found what I was after. These elliptic curves also have zeta functions with a palindromic symmetry. When we put the two together, suddenly, as if by magic, Christopher's calculations matched up perfectly. The palindrome was still there!

It was a really exciting moment. I felt a fatherly pride in Christopher for his achievement. He was experiencing the elation of that first taste of discovery, that flash of illumination that all mathematicians crave. If my crazy elliptic curve example still had this palindromic symmetry, there was renewed hope that my Palindrome Conjecture might still be true. I'd really expected Christopher's calculation to destroy the conjecture – after all, it had dramatically changed my perspective on the PORC Conjecture. But it had ended up doing the opposite. This new discovery spurred me on to understand why this symmetry was there, to prove my conjecture.

That day was memorable for another reason. Shani phoned me a few minutes after we'd found the new palindrome and said I should switch on the TV. Two planes had just smashed into the World Trade Center in New York. The date was September the 11th, 2001. Christopher and I watched for the next hour as the twin towers crumbled to the ground.

I went back to a theoretical description of the formula for the zeta function that Fritz and I had discovered a few years earlier. I was convinced that the secret of the palindromic symmetry was hidden somewhere inside it. But it still wasn't clear why this formula should always produce symmetry in the zeta function. I filled pages and pages of yellow legal pads, twisting and turning this formula, rewriting it in different ways, trying to find some way to rearrange it to see the symmetry. It was as if the formula was a Rubik's cube – I kept manipulating it, rotating its sides, in the hope that suddenly all the colours would match up to reveal the pattern I was convinced must be there. But nothing gave.

A few weeks after Christopher's breakthrough I phoned Fritz in Bonn. I'd sent Christopher off to the Max Planck Institute to show his work to Fritz. 'I think Christopher and I should be able to prove the functional equation by next week,' said Fritz. I remember going into a panic. I should have been happy, but I was devastated. This was what *I* wanted to prove. I knew that I might have been making a mistake in sharing my ideas, but I was being altruistic and thinking of the greater good of the subject. Now, I couldn't bear the thought that it might be proved by someone else. Had I been cast aside so early? I wanted to leap on a plane immediately. I felt like Tartaglia, letting his secret for solving cubics out of the bag only to see the young upstart Ferrari rush off and solve the quartic.

When Christopher got back from Bonn it turned out that their initial optimism had been misplaced. I hate to admit it, but I breathed a sigh of relief. I was back in the game. When I told some of my colleagues about this, they laughed nervously in recognition. There is a strange tension between the pride in one's students and the fear that they might kill you off.

A few years later things took another unexpected turn. I had arranged to meet my graduate student Luke in my office in Oxford. Luke is a wizard at programming. Previous calculations had been done in an ad hoc manner, requiring paper and pencil analysis of the twists and turns of each individual group of symmetries. Luke has found a way to get the computer to explore a group of symmetries in a much more systematic fashion, and if the group is straightforward enough, the computer identifies the ingredients that you need to calculate explicitly the formula for the zeta function. The calculations that Luke has started to produce have pushed our examples well beyond anything that we could do by hand. But it isn't simply a matter of cranking a handle – the computer is only as good as the person programming it. He still needs to guide the computer through the calculation.

Luke started to show me some of his latest calculations. He had a pile of papers with massive equations on them, some taking up several pages. Despite their size, again and again they displayed this beautiful palindromic symmetry: halfway through a formula, the coefficients start to repeat themselves in reverse order. We'd almost got to the end when Luke pulled out a sheet of paper at the bottom of the pile. 'Oh, and I found this example.' I stared at the formula. Halfway through, instead of neatly reversing itself it started doing something completely different.

Bang! I'd been trying to prove this palindromic symmetry for years, and now this example blew my conjecture out of the water. This formula was not a palindrome. This is where a computer comes into its own: it may not be very good at proving theorems, but it can be very effective at smashing conjectures. And now I understood why Luke had seemed so nervous when he first sat down. Partly it was excitement at having made such an important discovery, but he must have been wondering how I would react to being told that I'd spent ten years chasing a shadow. This was a far deadlier blow than having someone else prove my conjecture in front of my eyes. I really did have Oedipus sitting here in my office.

But I just had to swallow it. This is the set of cards that my subject has dealt me. What can I do? I can't rewrite the facts to suit my view of the world. There is something quite unforgiving about mathematical logic. This symmetry was so beautiful that I felt it had to be true, but logic was telling me otherwise. Mathematicians are driven in their work by a strong sense of aesthetics, and the correct path to follow is quite often the most beautiful one. And just as symmetry in nature signifies meaning, I had felt strongly that this palindromic symmetry contained a message about some internal structure that was making my zeta functions tick. There is no denying that many symmetry groups have zeta functions with this palindromic symmetry. So the big question now is this: why do some groups of symmetry give rise to palindromes, while others, like Luke's example, do not?

Luke's example also illustrates the power of the computer in mathematical research. Last week the newspapers reported the discovery of a prime number with over nine million digits; only the computer can give us access to such huge primes – and generate more data than we could previously have hoped for. If I had a huge amount of time to spare, perhaps I could have gone through the details of calculating Luke's example by hand. But undoubtedly I would have made a mistake. I certainly would have guessed that I'd made a mistake once I'd found that the answer didn't have the palindromic symmetry I was expecting. I'd often used this as a way of checking for errors in the past. It was very effective: once I got the symmetry, I knew the calculation was probably right because symmetry is hard to find by accident.

But as I'd discovered last month in Bonn, there are limits also to the computer's capabilities. Luke says that he is reaching the limits of the examples the computer can handle. Beyond these examples lies an infinite expanse of uncharted territory, navigable only by the human mind.

Luke's example will always be a warning to me to go into every problem with an open mind. His group of symmetries revealed that day a new subtlety in the theory of symmetry: things aren't monochromatic – there is a texture we were unaware of. We still haven't christened this phenomenon, but it deserves a name. Perhaps we should call groups of symmetries whose zeta functions have this symmetry 'palindromic groups'.

In my office that day I witnessed a new beginning. The subject had changed, and an addition to the language was required to describe this new vision. Luke is coming round again this morning, and we'll see whether his computer explorations have thrown up any other surprises.

It is one of the excitements of doing mathematics that a student might walk into your office, or a letter might arrive from some far-flung corner of the globe, with a breakthrough that changes the whole complexion of the subject. Surprises were in store for the mathematicians who picked up from where Tartaglia, Cardano and Ferrari left off, and were trying to cook up a formula to solve quintic equations. But they came from a rather unexpected source.

A glimpse through the Strait of Magellan

At the beginning of the nineteenth century, Norway was cut off more than it usually is. In winter the fjords often froze over, preventing access from outside the country, but in 1807 Norway, then still under Danish control, was also suffering from political isolation. That year Britain attacked the Danish fleet in a pre-emptive strike, fearing that it would be used by Napoleon to mount an invasion of British shores. Continental powers blockaded Britain in retaliation for this act of aggression. Britain saw a blockade of Norway as a way to punish the Danes. The blockade was devastating for Norway, because the country's principal export was timber to Britain. And the blockade starved the economy of vital funds; worse, the population was starved of food because the blockade cut off the grain supply from Denmark that Norway replied upon. By 1813, famine was rife in Norway.

In the midst of all the political isolation and turmoil this country was suffering, a young Norwegian was taking his first steps along the road to a mathematical career that would mark him out as one of the country's foremost mathematicians. With little contact beyond the few books that made it to Norway, Niels Abel would go on to solve a problem that centuries of work by eminent mathematicians in France, Italy and Germany had failed to crack. Renaissance mathematicians had unlocked the secrets of solving cubic and quartic equations. But for 250 years European mathematicians could not seem to find a way through the intellectual blockade that seemed to surround the question of solving the quintic until Abel finally found a way through.

Abel's interest in science had been sparked by his father, who would drag him out of bed in the middle of the night to witness a lunar eclipse or a passing comet. The stars thrilled the young Abel. Mathematics, on the other hand, didn't inspire him. Perhaps this was because his schoolteacher often resorted to violence to get his children to learn their multiplication tables. On one occasion the teacher got so carried away that he beat one of his students to death, for which he was expelled from the school. His place was taken by a young mathematician, Bernt Holmboe.

Holmboe was familiar with the exciting new mathematical developments taking place across Europe. His stories of great breakthroughs and outstanding challenges transformed the subject for Abel, who started to read some of the great works by Leonhard Euler and Isaac Newton. Within a year of Holmboe's arrival at the school, Abel was streaking ahead in his own mathematical development.

Abel's graduation from school was marred by the tragic death of his father, who died after years of excessive drinking. His father's alcoholism had been exacerbated by his humiliating dismissal from a high-ranking political position after he made false accusations against a political adversary. The 18-year-old Abel was left having to care for five children and an alcoholic mother who became a social outcast after being caught in bed with a lover on the afternoon of her husband's funeral.

As Abel's mentor, Holmboe was determined that the young man's fantastic mathematical talent should not be squandered, and from his own meagre income he paid for Abel to attend the newly opened university in Norway's capital. Abel was only too happy to escape the pressures of his family life by burying himself in the emotionally neutral world of mathematics. One of the great unsolved problems that particularly appealed to Abel was trying to find a formula that would give solutions to quintic equations, ones with terms in x^5. He'd read about the formulae discovered by Tartaglia, Cardano and Ferrari. These formulae seemed almost magical in their ability to produce the solutions to any cubic or quartic equation. He was determined to add his name to the list of eminent mathematicians who had cracked the secrets of equations by finding a formula to solve the quintic.

It was the sort of problem that even a naive mathematician can start to play with and try to construct a formula to do the job. A German mathematician, Walther von Tschirnhaus, thought he'd cracked the quintic in 1683, but his compatriot Gottfried Leibniz, Newton's mathematical rival, found mistakes in Tschirnhaus's analysis. As one seventeenth-century commentator, Jean-Étienne Montucla, wrote, 'the ramparts are raised all around but, enclosed in its last redoubt, the problem defends itself desperately. Who will be the fortunate genius who will lead the assault upon it or force it to capitulate?'

In its day, solving the quintic had the same status as trying to prove Fermat's Last Theorem (a problem Abel also had a go at) or cracking the enigma of the primes – two problems that have probably elicited the most 'solutions' from amateur mathematicians. Just as today I get several letters a week describing some theory about the origin of prime numbers, so too the professional mathematicians at the leading academies across Europe would hear from people claiming to have cracked the secret of the quintic. After a few months of tackling the quintic, Abel came to Holmboe with a formula that he believed would unravel this great unsolved problem of the age. Holmboe was amazed, but unable to check whether Abel's formula really did what he claimed. They sent it instead to the Danish mathematician Ferdinand Degen, for publication by the Royal Society of Copenhagen.

Degen wrote back asking to see the formula applied to an example. After all, this was the ultimate test for Abel's theory. It had to be able to solve any equation involving x^5. But when Abel sat down to show how his formula could be used to find the solutions of a particular quintic, he suddenly realized that there was a flaw in his workings. Maybe Degen had seen the same flaw but had very generously given Abel the chance to find it for himself. Degen obviously recognized the talent in Abel's letter. He even sent the young mathematician some interesting problems to try out, suggesting that perhaps he had the talents to 'discover a Strait of Magellan leading into wide expanses of a tremendous analytic ocean'.

Coming to understand the limitations of his formula had a striking effect on Abel. By going through the anguish of seeing his great discovery crumble before his eyes, he gained a deeper insight into the subtleties of the quintic equation. It completely changed his perspective on the problem. He began to see how he could prove why there might not actually be a formula to solve these equations. Almost like an initiation rite, Abel's baptism by equations marked his transition from an amateur calculator to a mathematician who would eventually be recognized as one of the greatest of his age. But that recognition still had to be earned, and that would result in Abel paying the ultimate price.

Thanks to a small grant from the university, Abel got the chance to visit Degen in Copenhagen to discuss his mathematics. He could sense that Norway was too isolated and lacked the stimulating environment that he needed if his ideas were to be brought to fruition. In fact Copenhagen lacked the kind of mathematical stimulation he craved, but his visit did result in him meeting and falling in love with a young woman, Christine Kemp. Abel described her to a friend as 'not beautiful; she has red hair and freckles, but she is an admirable woman'. Abel realized that he was not yet sufficiently well established to marry, but he promised her that once he had got the professorship in Norway that he coveted and was secure in his position, he would marry her. She agreed to wait for him to make his name.

Spurred on now by the promise of what the future might hold, Abel dedicated his efforts to trying to understand why these quintic equations seemed so uncrackable.

Dismissing mental tortures

By this time, mathematicians had begun to understand the numbers that Cardano had dismissed as 'mental tortures'. The square roots of negative numbers that had appeared in Cardano's calculations were not 'so subtle that they are useless', as Cardano had thought. By the beginning of the nineteenth century these 'imaginary numbers', as René Descartes had first called them, were finally being accepted as an integral part of the world of mathematics. Mathematicians accepted that these numbers had been there all along, and it was up to them to uncover their secrets. And in doing so they started to discover why the world of symmetry might be intimately linked to solving equations.

The history of mathematics shows how different cultures wrestled with the discovery of new types of number. The concept of a whole number is part of our evolutionary make-up. Our brains appear to be hard-wired to identify 1, 2, 3, ... Indeed, mathematicians call them the natural numbers in recognition of their fundamental place in the natural world. The ability to recognize the 'two-ness' of something as different from 'three-ness' is what led peoples the world over to introduce words and symbols to articulate this difference.

The fact that our brains have evolved to recognize the concept of whole numbers can probably be attributed to the Darwinian principle of survival of the fittest. The ability to assess how many animals there are in the opposing pack will inform the decision to fight or fly. Research has shown that animals are not only able to compare but can also count. Monkeys and cats count their young to check they are all there; coots know when the number of eggs in their nest has increased, indicating that another bird has sneaked in a parasite egg; human babies as young as five months can tell when one of their dolls is taken away from a pile; even dogs seem to twig that something fishy is going on when experimenters try to trick them into thinking that $1 + 1 = 3$.

But humans have taken these numerical foundations and built new types of number from the basic 1, 2, 3, . . . Each new type, such as negative numbers or imaginary numbers, seemed totally unnatural at first, and it often took generations for scientists to accept them and take them on board. But gradually, as people produced graphical representations of these numbers, or rules and equations for how to manipulate them, a language emerged which allowed the next generation to talk with confidence about these discoveries. Those who pioneered solving quadratic equations found the going harder because negative numbers were not part of the lexicon of the day.

The problem of solving equations led to the discovery of a host of new types of number. Negative numbers arose as solutions to questions such as 'find x such that $x + 3 = 1$'. Fractions help us to divide numbers that don't naturally divide: for example, the question of how to divide seven loaves of bread between three people, captured in the equation $3x = 7$, needs the language of fractions to enable us to talk about something less than a whole loaf.

Pythagoras and his followers had believed that all mathematical problems could be solved using whole numbers and fractions created from whole numbers. It therefore came as a shock to discover that the length of the long side of a right-angled triangle whose two short sides are one unit long could not be expressed as a fraction. Pythagoras's Theorem implied that the length was a number whose square is 2 (Figure 43). What was this number? If you take the fraction $7/5$ then

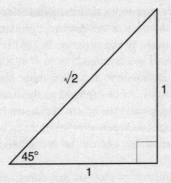

Fig. 43 A triangle whose longest side is the square root of 2 in length.

its square is quite close to 2; 707/500 squared gets you closer. Pairs of larger and larger numbers could be found whose ratio was closer and closer to the length of the side of the triangle, but, the Pythagoreans proved, no fraction captured the length exactly.

The square root of 2 and other numbers that cannot be expressed as fractions are called irrational numbers, meaning numbers that cannot be expressed as ratios between whole numbers. More irrational numbers arose out of Tartaglia's equations. He needed to take cube roots of numbers to solve equations such as $x^3 = 2$. The cube root of 2 is the length on the side of a cube enclosing a volume of size 2. But it was with some unease that mathematicians started to accept numbers such as the fifth root of 2. It was less clear what geometric meaning such a number might have – it seemed to require conceptualizing boxes in five dimensions. Omar Khayyam had dismissed such numbers as meaningless. Nevertheless, as mathematicians moved away from the geometric ties that characterized the mathematics of the ancient world and towards a more abstract view, they began to explore the arithmetic of numbers for their own sake. Although the fifth root of 2 didn't have any obvious geometric interpretation, they realized that there were numbers such as 11,487/10,000 which, when raised to the fifth power, could get you very close to 2.

Lengths of geometric objects continued to produce other new numbers. Take a length of string, and fix one end to the floor and attach a pen to the other end. Mark out a circle in the ground. If the distance across the circle is one unit, how far does the pen travel? This is such an important number that mathematicians gave it a special symbol: π. From the time of the Ancient Egyptians, mathematicians tried to write this quantity as a number. In 1761 it was proved that π cannot be represented as a fraction.

In 1882 it was shown that π is even more mysterious. It was a number that couldn't even be captured as the solution to one of the equations that Cardano and the others had shown how to solve. These numbers that cannot be expressed as solutions to equations had been named transcendental by Leibniz in recognition of their elusive character.

Despite the slippery nature of irrational and transcendental numbers, mathematicians began to build up a picture of where all these numbers 'were': fractions, irrational numbers such as the fifth root of 2 and important constants such as π could all be thought of as points sitting along a ruler or line, something we now call the number line. In principle, all these different numbers could be measured by a ruler. The physical reality of all whole numbers, fractions, and irrational and transcendental numbers led mathematicians to refer to them collectively as the real numbers.

The problem came when mathematicians tried to make sense of solving $x^2 = -1$. The Indian mathematicians had proved that a negative number squared gives a positive number. So there didn't seem to be any numbers on the number line that would solve this equation. There were two ways forward: either declare that these equations could not be solved, or – the more daring option – invent new numbers to solve these equations.

Ever since Cardano encountered the need for square roots of negative numbers, mathematicians had been moving closer and closer to taking the bold step of admitting these new numbers into the mathematical canon. But would this unleash too many new types of number? For example, what if we want to find a number whose fourth power is -1? Would that require the invention of yet another new number?

The amazing conjecture that mathematicians put forward in the seventeenth century was that just one new number, the one needed to solve $x^2 = -1$, could be combined with whatever was necessary from the realm of real numbers to produce a solution to any other equation. Not everyone was so confident that this would work. Leibniz was certainly of the belief that you'd need to invent more and more numbers. He couldn't see how introducing just the square root of −1 could help you answer the more complicated question of finding the fourth root of −1.

The new number whose square was −1 was called an imaginary number and given the symbol i. It took two hundred years to prove that this imaginary number was as powerful as people believed. Eventually it was Gauss who proved in his doctoral thesis that any equation of the form $x^6 + x^5 + 3 = 0$ or $x^4 = −1$ could be solved by setting x equal to a number $a + bi$ built out of this imaginary number i and two real numbers a and b. This combination $a + bi$ of imaginary and real numbers is called a complex number.

For example, setting a and b equal to $1/\sqrt{2}$ will give a number of the form $a + bi$, whose fourth power is −1. If you are brave enough to do some algebra here is why:

$$\left(\frac{1}{\sqrt{2}} + \frac{1}{\sqrt{2}} i \right)^2 = \tfrac{1}{2} + (2 \times \tfrac{1}{2} \times i) + \tfrac{1}{2} i^2 = i$$

and the i, when squared again, gives us −1.

So important was this discovery that it became known as the Fundamental Theorem of Algebra. In proving this result, Gauss answered another problem that had been bugging mathematicians: just where were these imaginary numbers, if they were not on the number line? He produced a picture of these new numbers which gave them some semblance of reality and started to hint at the connections that solving equations might have with symmetry.

Gauss had actually used a picture of the imaginary numbers as a mathematical tool in his proof, but he kept it hidden for many years, fearing he would be laughed at by a mathematical establishment still wedded to the language of equations and formulae. But because the image was so powerful and gave imaginary numbers a physical reality, it was only a matter of time before others hit upon the idea. Two amateur mathematicians, the Dane Caspar Wessel and the Swiss Jean Argand, independently proposed similar pictures in pamphlets they published. Argand, who was the last of the three to have the idea, is the person whose name became attached to the picture we now call the Argand diagram. Credit is rarely just.

The idea was that if there were no numbers on the number line

whose square is −1, why not create a new direction to represent these
new numbers whose squares were negative? A two-dimensional picture
emerged in which the regular number line became the horizontal axis,
representing the real numbers, and the vertical axis could be used
to record the imaginary numbers (Figure 44). So on the Argand diagram the complex number 3 + 4i, for example, is represented as the
point (3, 4).

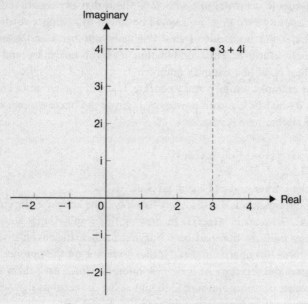

Fig. 44 The Argand diagram of complex numbers.

Once the picture of these complex numbers became known, mathematicians realized how powerful this representation could be. Adding
complex numbers was just like following two sets of directions, one
after the other. Mathematicians also discovered the beautiful fact that
multiplying numbers translated into rotating numbers around the
point representing the number 0 which on the Argand diagram is
called the origin. To find the direction of the product of two complex
numbers, you join the point representing each number to the origin,
then add the angles these two lines make with the horizontal axis.
Whereas, centuries before, algebra had thrown off its ties with geometry in order to develop, now progress was being made by moving in

the opposite direction. It was like having a dictionary which translated the algebra of these numbers into geometry – two different languages for the same thing. The power of this dictionary was that certain ideas were much more self-evident in one language than in the other.

For example, mathematicians discovered that the solution of the equation $x^4 = -1$ was a number which you got to on the Argand diagram by travelling a distance $1/\sqrt{2}$ horizontally, in the real direction, and the same distance vertically, in the imaginary direction (Figure 45). This number, $x = 1/\sqrt{2} + 1/\sqrt{2}i$, is actually sitting on a circle of radius 1. You can get to it by starting at the number 1 – the point $(1,0)$ – and then making an eighth of a turn anticlockwise round the circle.

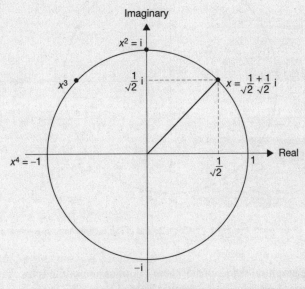

Fig. 45 The point x marks the location of the fourth root of −1 in the Argand diagram. Raising the number x to the power of 4 moves the point round the circle to −1.

There is a rather beautiful geometric way to raise this number to the power 4. Each time the number is multiplied by itself, the location of the result moves round the circle by an eighth of a turn. After four spins we get to the number −1, the point $(-1,0)$ on the Argand diagram. The geometric language turns out to be much simpler to play with than the algebra. Not only that, it also reveals that there

are actually three other numbers which are solutions of the equation
$x^4 = -1$. If you take any of the four points x, y, z or w on the circle and
spin them in a similar way, they all land on the number -1 (Figure
46). For example, the number y represents a spin of ⅜ of a turn from
the number 1 on the real axis. To locate y^4, repeat this turn four times
and you get to the number -1. So there are actually four complex
numbers that will solve this equation.

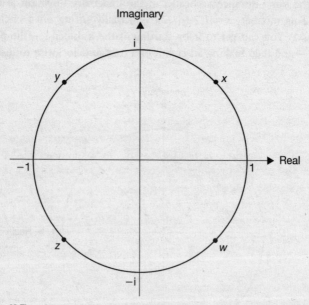

Fig. 46 The points x, y, z and w mark the location of four complex numbers whose fourth
power is -1.

The graphical depictions of these solutions and of the spins that get
you from one solution to another start to reveal the geometric symmetry that underlies an equation such as $x^4 = -1$. Articulating this
connection more explicitly would ultimately unlock both the secrets
of equations and a new language to describe symmetry.

The discovery that there were four different complex numbers that
could solve the equation $x^4 = -1$ was an important breakthrough. Omar
Khayyam's analysis of equations had allowed him to see that there
might be more than one answer to these equations. By 1629, the
Flemish mathematician Albert Girard proposed that the number of

solutions will depend on the highest power in the equation. So if the equation is a quartic, such as $x^4 = -1$, you will expect four solutions. The equation $x^7 = -1$ has seven solutions. The obvious one is got by setting $x = -1$, but there are another six which, together with $x = -1$, are evenly arranged round the unit circle on the points of a heptagon. We have actually met a simple example of this already, in that $x^2 = 4$ has two solutions: $x = 2$ and its mirror solution, $x = -2$. What mathematicians were now proposing was a much wider generalization: there are five fifth roots, seven seventh roots, and so on.

These extra solutions begin to show us how symmetry might be related to solving equations. Mathematicians would eventually discover that every equation has some symmetrical object attached to it. For the equation $x^4 = -1$ it is some (but not all) of the symmetries of the square connecting the solutions x, y, z and w.

Niels Abel was already beginning to see that symmetries in the solutions of these equations were the key to understanding whether the solutions to the quintic could be found by equations involving fifth roots, just as the cubic had been solved using an equation involving cube roots. Clearly an equation such as $x^5 = 3$ can be solved using fifth roots. But what about $x^5 + 6x + 3 = 0$? Tartaglia had found ways to twist the cubic until it looked like an equation that involved taking cube roots and square roots. Was the same trick possible for the quintic? By 1824 Abel had cracked it. There was something about the symmetries associated with the five solutions of the quintic which meant that there was no formula for the solution of any quintic equation. Finally, Abel understood why the formula he had sent to Degen three years before was ultimately doomed to fail.

Showing that no formula exists for the quintic was a problem of a very different order of complexity to the one tackled by Tartaglia, Ferrari and Cardano. How on earth could you really convince people that despite all leaps of the imagination you could never find a formula to solve the quintic? The formula produced by Tartaglia was a tangible thing that you could check. Now the quintic was forcing mathematicians into having to think much more conceptually. With this move to abstraction came the seed for a language of symmetry.

Shuffling solutions

Abel's genius was to break with the shackles of the past with all its explicit equations and formulae and apply a more abstract theoretical analysis to these algebraic expressions. In school we are taught al-Khwārizmī's formula for solving quadratic equations. If you want to find the solutions to the quadratic equation $ax^2 + bx + c = 0$, then feed the numbers a, b and c into the formula

$$x = \frac{-b \pm \sqrt{(b^2 - 4ac)}}{2a}$$

Abel proved that when it came to quintic equations, however complicated a concoction of square roots, cube roots or higher roots you cooked up for your formula, there would be a quintic equation for which your formula would fail to find the solutions. To prove his claim he made one of the classic moves in the mathematician's gameplay: assume that there *is* such a formula, and then show why that must lead to a contradiction.

Reductio ad absurdum is the name of this move – keep making deductions from a hypothesis until you get something absurd. Then you can conclude that the hypothesis must have been false. It was a trick the Pythagoreans were adept at; it was how they proved that you couldn't represent the square root of 2 as a fraction. If you suppose that you could, you end up proving that an even number is equal to an odd number. The famous Cambridge mathematician G. H. Hardy wrote in his book *A Mathematician's Apology* that *reductio ad absurdum* was one of mathematics' finest weapons: 'It is a far finer gambit than any chess gambit: a chess player may offer the sacrifice of a pawn or even a piece, but a mathematician offers the game.'

Having assumed that the quintic could be solved with a formula that would tell you what the five solutions were, Abel began to play around with what that might imply, hunting for some ultimate absurdity that would kill his hypothesis. He knew that a quintic equation of the form $x^5 - 6x + 3 = 0$ has five different solutions, which we shall call A, B, C, D and E. These letters are names for the five numbers that solve the equation. Abel could also think of these letters as labelling five points on the map of complex numbers. He then began to consider all the *new* formulae that you could build from the five solutions, formulae such as $B - A \times C - D \times E$ or $A \times B \times C \times D \times E$.

His inspirational step was to consider what happens when you take a formula and swap the five solutions around. For example, what happens to the answer to the formula $A \times B \times C \times D \times E$ if you change the order of the individual solutions? In this case, nothing. The formula $A \times B \times C \times D \times E$ multiplies all the solutions together and doesn't yield a different answer if I swap some of the solutions around and calculate $B \times A \times C \times D \times E$ instead. But a formula such as $A - B \times C - D \times E$ in general will give a different answer when I swap A and B and calculate $B - A \times C - D \times E$ instead.

What is the maximum number of different answers one formula could give by swapping round the numbers A, B, C, D and E? I need to calculate how many different ways I could arrange the five letters in this formula. Here is one possible rearrangement:

A	B	C	D	E
↓	↓	↓	↓	↓
D	A	E	B	C

Counting all the possibilities is a little like working out how many different combinations there are on a lock with five wheels where each wheel has the letters A to E engraved on it, except that now the combination lock is not allowed to show the same letter twice. Another way to think of this is to count how many five-letter 'words' you can make using the letters A to E once and once only. There are five choices for the first letter. In the above rearrangement, we chose the letter D. Having chosen the first letter we are now only left with four choices for the second letter, then three for the third, two for the penultimate letter and no choice at all for the last one. So there are $5 \times 4 \times 3 \times 2 \times 1 = 120$ different words we could build. Each word represents a different arrangement of the letters in the formula we are looking at. And with 120 different ways to order the five solutions A, B, C, D and E, there are potentially a maximum of 120 different answers.

But it depends on the formula. Some formulae give just one answer regardless of the order of the numbers. For example, $A \times B \times C \times D \times E$ doesn't change its value if you swap any of the letters around. Some

formulae take two values. For example, swap the letters around in the
formula

$$(A-B)(A-C)(A-D)(A-E)(B-C)(B-D)(B-E)(C-D)(C-E)(D-E)$$

and you get either the number you started with or its negative: two
possible answers. Abel's crucial breakthrough was to prove that you
could never build a formula out of these five solutions, A, B, C, D
and E, which generates only three different answers for all possible
permutations of the numbers. It was also impossible to get just four
answers out of a formula. But there was a formula that generated at
least five different answers. A gap had opened: you jumped from
formulae giving two answers to formulae giving five answers.

Formulae built from solutions to quadratic or cubic or quartic
equations didn't have this gap. It only started to appear when the five
numbers A, B, C, D and E were solutions to a quintic. But how could
Abel exploit this realization?

He began by assuming that there was some magic formula for these
solutions A, B, C, D and E like the one the Arabs derived for quadratic
equations or Tartaglia's formula for cubics. Having made this assump-
tion, he began to look for a contradiction. Abel first deduced a general
expression for what this formula would look like if it existed. He then
took this hypothetical magic formula for the solutions A, B, C, D and
E and plugged it into the formulae such as $A - B \times C - D \times E$ that he
had been exploring, to see if it led to any absurdity or contradiction.
After pages of twisting and turning, Abel finally teased out why a magic
formula would contradict the fact he'd proved earlier: that a formula
in the solutions A, B, C, D and E can't take three or four answers. This
contradiction was enough to show that the magic formula therefore
could not possibly exist.

What Abel didn't appreciate was that by thinking of A, B, C, D and
E as points on the map of complex numbers, the permutation of the
five numbers would look more like geometric moves which permute
these points around. For example, if the solutions were arranged evenly
round a circle then the permutation that sends A to B, B to C, C to D,
D to E and E to A is the same as a rotation of the circle through a fifth
of a turn. The language Abel was beginning to formulate to express
his ideas about equations would ultimately help to articulate symmetry.

The 120 words built from the letters A, B, C, D and E are the beginning of a language to describe the geometric moves that permute points on the map. But Abel was more interested in the fact that he'd just cracked one of the greatest problems of his age: there was no magic formula to solve quintic equations.

The cantankerous Cauchy

Abel knew that he had in his hands the passport to the academies of France, Italy and Germany. Here was his chance to make his name, get that professorship in his native Norway and finally marry the woman he had fallen in love with in Copenhagen. Despite extreme poverty, he got together enough funds to publish his solution himself. He could afford only enough paper for six printed pages. It had taken him pages and pages of handwritten notes to arrive at his contradiction. But faced with the costs the printer was demanding, he managed to distil the key points of his argument to fit the limited space he had available. The ideas were tough enough without being stripped to the bare bones, but Abel had no choice. The paper opened with these words:

> Mathematicians have occupied themselves a great deal with the general solution of algebraic equations and several among them have sought to prove the impossibility. But, if I am not mistaken, they have not succeeded up to the present. Therefore I hope that mathematicians will receive kindly this memoir, which aims to fill this gap . . .

It then leapt straight into the logical onslaught of Abel's ideas.

Abel sent his paper to the leading mathematicians of the day, but it was ultimately the French Academy he hoped to impress with his proof. Founded in 1666, the Academy had grown from a few scientists meeting in the Royal Library to an institution driving the scientific agenda in Europe. The prizes it instigated in 1721, awarded for solutions to particular problems, had become so influential that they defined the future of science in the ensuing decades. The first prizes were awarded for problems about masts of ships, charting the stars at sea and memoirs on the compass – problems clearly driven by practical considerations. But as the century progressed, the problems began to reflect an interest in more abstract mathematical concerns.

The French Academy met regularly to present and discuss the outstanding achievements of the day. The academicians' seal of approval was essential for any budding mathematician, which is why Abel prepared his paper on the quintic equation in French. The mathematician whom Abel really hoped to impress was the academician Augustin-Louis Cauchy. If he could only get Cauchy to present his paper to the Academy, he knew that he would no longer be standing out in the cold, peering in through the windows of the mathematical drawing rooms of Europe. He would be welcomed inside as the new champion of the age. Cauchy had a reputation in the Academy as one of the toughest nuts to crack. Whether Abel knew it or not, Cauchy had been working on questions of symmetry and solving equations but had not drawn any connection between the two.

Cauchy was born in Paris on 21 August 1789, just over a month after the storming of the Bastille, and had suffered a particularly harsh childhood. His father had been an aspiring part of the regime which the revolutionaries were determined to overthrow. As the Reign of Terror in Paris mounted, Cauchy's father became more and more worried as he saw friends and compatriots tried and executed. The family decided to flee to the refuge of the country. Life was still not easy. Cauchy's father wrote of their hardship that 'We never have more than a half pound of bread – and sometimes not even that. This we supplement with the little supply of hard crackers and rice that we are allotted.'

Cauchy contracted smallpox, which severely weakened him. The constant threat hanging over the family of the arrest of his father and the exhaustion caused by starvation probably contributed to the development of Cauchy's introspective character. He was rarely seen playing with other children of his age; instead, he sought solace from the turmoil in his family's life by escaping into his books. Although he loved languages and literature, it was becoming clear that mathematics held a particular fascination for him. His teacher noted that 'it was not an infrequent thing to find a paper on a literature assignment suddenly interrupted. A mathematical idea would have crossed the youngster's mind and so absorbed him that he would be forced to translate the compelling notion into numbers and figures.'

When the Terror came to an end the family moved back to Paris, and Cauchy's father resumed his upwardly mobile political life. Eventually his hard work was rewarded when he was elected to the Senate. There he became friends with two fellow senators who were also leading mathematicians: Pierre-Simon Laplace and Joseph-Louis Lagrange. Lagrange had become famous before the Revolution for solving some of the Academy's prize problems, including work on the moons of Jupiter and Saturn, the influence of the planets on passing comets and why the Earth's moon oscillates slightly, showing different features of the lunar surface. He credited his father's bankruptcy for inspiring his choice to become a mathematician in the first place: 'if I had been rich I probably wouldn't have devoted myself to mathematics'.

But like Cauchy's father, Lagrange was extremely fearful when the Revolution swept through France. He survived – but only just. Because he was born in Italy he faced arrest under the new laws of September 1793. However, the eminent chemist Antoine Lavoisier argued Lagrange's case, and the mathematician's name was specifically excluded from the edict. A year later, Lavoisier was sent to the guillotine; as a former collector of taxes for the deposed monarch, he stood no chance of being pardoned by the revolutionaries. 'It took only a moment to cause this head to fall, and a hundred years will not suffice to produce its like,' wrote Lagrange of the friend who had saved him from arrest. Some credit Lagrange's survival during this period to his public views that 'one of the first principles of every wise man is to conform strictly to the laws of the country in which he is living, even when they are unreasonable'. Perhaps this was a lesson he learnt from the unforgiving world of mathematics.

One day, Cauchy's father let his son accompany him to work at the Palais du Luxembourg, and he showed the great academicians some of the mathematics that his son had been working on. Lagrange was impressed and, turning to his colleague Laplace, declared: 'You see that little young man? Well! He will supplant all of us in so far as we are mathematicians.' Lagrange's advice to Cauchy's father concerning the boy's education was unusual: 'Don't let him touch a mathematical book nor write a single number before he has completed his studies in literature.' His linguistic skills should be developed first, said Lagrange.

Lagrange could already sense important changes afoot in mathematics. There was a need for a more abstract language, sophisticated enough to express the subtleties of this new mathematical age. Perhaps if the young Cauchy were well grounded in the grammar and rules of Greek and Latin, he would have the foundations for creating this new abstract language of mathematics. 'If you don't hasten to give Augustin a solid literary education,' warned Lagrange, 'his tastes will carry him away; he will be a great mathematician but he won't know how to write his own language.'

Cauchy entered university at the precocious age of 16, but despite the academic environment he still felt an outsider. His fellow students were all fired up with the revolutionary politics of the day. Cauchy, on the other hand, had inherited his father's staunch Royalist views and his mother's pious Catholicism. His fellow students taunted him mercilessly for his overt political and religious opinions. Cauchy stuck to his strongly held beliefs, and even joined a secret Catholic society that sought to install those with allegiances to the Pope in positions of influence.

Despite Cauchy's opposition to the Republican cause, Napoleon enlisted his services to join the engineers in Cherbourg who were building a fleet to invade England. Following his graduation, Cauchy laboured away for three years: 'I get up at four and am busy from morning to night. Work doesn't tire me; on the contrary it strengthens me and I am in perfect health.'

Lagrange remained interested in the mathematical prodigy and suggested that he might like to look at a problem that was perplexing mathematicians at the time. It concerned certain new symmetrical shapes that had been discovered. Plato's colleague Theaetetus had proved two thousand years before that there were five Platonic solids – three-dimensional objects whose faces were all copies of the same regular polygon. For example, 12 pentagons could be put together to make the 'sphere of pentagons' or dodecahedron.

To everyone's surprise, in 1809 a new shape had been built out of these 12 pentagons. Theaetetus had insisted that the faces of his shapes should not cut into each other. But what if you relaxed this condition? A mathematics teacher in Paris had found a new way to piece 12 pentagons together to make a new symmetrical shape that was christened the great dodecahedron (Figure 47). Although it looks like a shape built from lots of irregular triangles, it consists of 12 intersecting pentagons. The shape satisfies all the conditions for a Platonic solid except for the fact that the faces cut into each other. How many other strange and beautiful shapes like this might be out there? Three others were soon discovered, and mathematicians began to wonder where the new list might end.

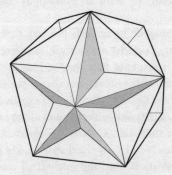

Fig. 47 The great dodecahedron.

The French Academy decided that the issue needed to be settled, and dedicated its prize for 1811 to the question of proving beyond doubt that the five Platonic solids plus the four new solids were all the three-dimensional shapes that you could build from identical regular polygons. While he was hard at work as an engineer in Cherbourg, preparing Napoleon's fleet for the invasion of England, Cauchy set his mind to seeing whether these four shapes were the only additions to the symmetrical shapes of the Greeks. If he were to claim the prize on offer, he would require a watertight argument to prove why there couldn't be any more. Constructing new shapes was fine. Their existence, once built, spoke for itself, rather like the formulae that solved the quadratic, cubic and quartic equations. But it required very sound logical argument to convince the scientific establishment that there was no other sneaky way to piece together shapes to make a new object.

Cauchy was starting to experience the challenge of finding a language to articulate the visual world of geometry and space. As Descartes had declared, 'Sense perceptions are sense deceptions.' The way he tackled the problem marks a turning point in the way mathematics was being done. He recognized the weakness of appealing to one's geometric intuition and sought instead a more rigorous way to express intuitive ideas which might avoid the pitfalls of visual deceptions. This contrasted with the attitude of Renaissance scientists such as Johannes Kepler, who several centuries earlier had heaped scorn on the idea of turning pictures into language: 'nothing is proved by symbols, nothing hidden is discovered in natural philosophy through geometric symbols'.

Armed with his burgeoning critical approach to the mathematics of space and symmetry, Cauchy successfully answered the Academy's question for 1811. These four new shapes, plus the five classical Platonic shapes, were the only symmetrical shapes possible. The Academy awarded him its prize. But the effort he had expended in trying to get his head around shapes as well as the rigours of the engineering project at Cherbourg took their toll. Cauchy collapsed in September 1812 from severe depression and mental exhaustion. He returned to Paris, realizing that his isolation from the intellectual centre had not been good for him. Paris was where the action was and where he should be doing his mathematical work. Having won the Academy's prize, he soon assumed his place alongside the academic elite.

Ruffini's tiny mistake

Despite the appearance of these new symmetrical shapes, mathematics was still lacking any coherent theory of what symmetry really was. How could you say that two objects had the same or different symmetry? There was still too much emphasis on the physical reality of these objects as opposed to a theoretical understanding of the essence of what made them symmetrical or not.

The text that Abel had sent to Cauchy contained the beginnings of a language for symmetry. But it wasn't the only text that Cauchy had received about the quintic. In contrast to Abel's modest six-page pamphlet was a 500-page treatise by the Italian doctor and mathematician Paolo Ruffini that Cauchy had waded through several years earlier. Ruffini was also claiming to have proved that there was no formula to solve the quintic, but unlike the poverty-stricken Abel, he could afford paper – lots of it.

It was Lagrange who had introduced Cauchy to this huge opus. Lagrange didn't think it was worth much, but Cauchy with his youthful enthusiasm set out to work his way through it. Ruffini's work was inspired by a paper Lagrange had written 30 years before. Mathematicians were still exceedingly reluctant to admit that there was no formula to crack the quintic, which is why Ruffini, the first person to grab the nettle, has received scant recognition for his brave step.

Ruffini was convinced that his breakthrough would make him famous. He had cracked, in Lagrange's words, 'the most celebrated and important problem of algebra'. He decided to send his paper to the 'immortal' Lagrange, who had inspired his proof. Being a fellow Italian, Ruffini felt sure it would be well received. But far from the accolades Ruffini was expecting, his paper seemed to make no waves at all. He got no reply from Lagrange. He decided to write a second version, which he sent to Lagrange in 1801 with a desperate letter begging for some recognition even if it were negative:

> Because of the uncertainty that you may have received my book, I send you another copy. If I have erred in any proof, or if I have said something which I believed new, and which is in reality not new, finally if I have written a useless book, I pray you point it out to me sincerely.

Still no response. In 1802 Ruffini sent another version: 'No one has more right . . . to receive the book which I take the liberty of sending you.'

Fellow compatriots were supportive, but their backing was based more on partisan motives than hard, cold analysis. 'I rejoice exceedingly with you and with our Italy, which has seen a theory born and perfected and to which other nations have contributed little,' wrote a professor in Pisa to Ruffini on receiving his manuscript.

But there was a crucial problem with Ruffini's text. He had made a mistake. If only someone had pointed it out to him, he might have been able to correct it and claim the credit he was due. He had fatally assumed something rather special about the magic formula that he wanted to show did not exist. But he offered no argument for why one could assume that the formula, if it existed, had this special property. It was a missing piece in the jigsaw of the proof. Without it, it was as useless as Ruffini claiming to be descended from Julius Caesar with the exception of one gap in the family tree.

With the failure of the mathematical establishment to recognize his work, Ruffini returned instead to his medical practice. While treating patients in a typhus epidemic that swept Italy in 1817, he caught the disease himself. He never fully recovered, but did manage to publish a memoir about his experiences of the disease. A few months before he died, in 1822, he received a letter from Cauchy. His work on the quintic hadn't gone completely unrecognized:

> Your memoir on the general resolution of equations is a work which has always seemed to me worthy of the attention of mathematicians and which, in my judgement, proves completely the impossibility of solving algebraically equations of higher than the fourth degree.

He had not yet picked up the subtle error Ruffini had made.

Cauchy's praise for Ruffini was somewhat out of character. The general perception of Cauchy is that of a self-obsessed mathematician interested only in his own discoveries, never keen to give credit where credit is due. This side of his character emerged when he came to describe Ruffini's ideas to the Academy. Rather than presenting Ruffini's work at the Academy's weekly meeting, he presented instead his own generalization of Ruffini's result. Ruffini's original result, which had inspired Cauchy, didn't get a mention.

It would not be the last time that Cauchy put self-promotion ahead of recognizing those who had laid the groundwork. His successes were beginning to breed in him a rather arrogant streak that would become increasingly off-putting to his contemporaries. Jean-Victor Poncelet, a colleague of Cauchy's in Paris, describes Cauchy brushing him aside when they met in the streets in Paris. He had just received a letter from Cauchy rejecting his work for presentation to the Academy:

> I managed to approach my too rigid judge at his residence ... just as he was leaving ... During this very short and very rapid walk, I quickly perceived that I had in no way earned his regards or his respect as a scientist ... without allowing me to say anything else, he abruptly walked off, referring me to the forthcoming publication of his *Leçons*

l'École Polytechnique where, according to him, 'the question would be very properly explored'.

His colleagues had by now formed a very negative view of Cauchy. Even his mathematics they claimed was more negative than positive: he had won his prize at the Academy, not for constructing a new symmetrical object but for showing there were no new shapes to add to the list:

> He has introduced into science only negative doctrines ... it is in fact almost always the negative aspect of the truth which he came to discover, that he takes care to make evident: if he had found gold in whiting, he would have announced to the world that chalk is not *exclusively* formed of carbonate of lime.

'Yours destroyed'

Given the political and geographical isolation of Norway at the beginning of the nineteenth century, and the poor reception of Ruffini's work in Paris, it is likely Abel would have been unaware of the Italian's progress. While Ruffini's work covered hundreds of pages and contained a mistake, Abel had condensed his proof onto six sides. More importantly, Abel had not made the mistake in his argument that so fatally damaged Ruffini's claim to have cracked the enigma of the quintic.

In the autumn of 1825, Abel set off with four friends for his grand tour of Europe, hoping that the paper he had sent on before him would assure him a welcome reception. The trip was quite daunting: Abel was only 23 years old, and it was his first venture so far afield. The grant he had received to fund his trip had stipulated that he spend as much time as possible in Paris, given that the city was the Mecca of the mathematical world. His friends planned to visit Germany first. Abel was reluctant to travel to France on his own. He wrote back home: 'Now I am so constituted that I cannot endure solitude. Alone, I am depressed, I get cantankerous, and I have little inclination to work.' He decided to join his friends on their visit to Berlin.

In Berlin he made friends with a dynamic civil servant working at the Prussian Ministry for the Interior. August Crelle was passionate about mathematics. He organized soirées for young mathematicians to discuss their ideas, and set up a new mathematical journal in which he planned to publish the work of promising young mathematicians.

Crelle had a reputation for sniffing out great talent, and it didn't take him long to see that the young Norwegian who had been attending meetings at his house was exceptional. The first volume of Crelle's journal featured no fewer than seven papers by Abel, including his work on the quintic. Abel wrote back to his old teacher Holmboe:

> You cannot imagine what an excellent man [Crelle] is, exactly as one should be, thoughtful and yet not horribly polite like so many people, quite honest, for that matter. I am with him on as good terms as I am with you or other very good friends.

Abel felt sure the publication of his work would stand him in good stead in his application for the mathematics chair at the only university in Norway. The position would finally allow him to marry his fiancée. He was crushed to receive in Berlin a letter telling him that the professorship had been offered to, of all people, his old tutor Holmboe. Despite his love for his teacher, Abel knew that Holmboe's mathematics did not compare with his own work. And Holmboe was young enough that there was little chance of the chair becoming vacant again in the foreseeable future.

Nevertheless, Crelle's support gave Abel the psychological boost to make his way to Paris to find out what people had made of his work. He still procrastinated, making detours via Italy and Switzerland. 'My God! I, even I, have some taste for the beauties of nature, like everyone else. I shall make this one voyage in my life.' Finally he arrived in Paris excited by the prospect of meeting the great mathematical names of the age.

Much to his disappointment, no one seemed interested. Cauchy had failed to present any of Abel's papers to the Academy and seemed completely absorbed in his own work. Abel wrote to Holmboe:

> The French are much more reserved with strangers than the Germans. It is extremely difficult to gain their intimacy, and I do not dare to urge my pretensions as far as that; finally, every beginner had a great deal of

difficulty getting noticed here. I have just finished an extensive treatise on a certain class of transcendental functions to present to the Institute which will be done next Monday. I showed it to Mr Cauchy, but he scarcely deigned to glance at it.

Abel's funds by this time were running perilously low. He rationed himself to one meal a day. His evenings were spent playing billiards or sneaking into the theatre, which he loved, but the complete lack of interest in his work was beginning to grind him down: 'they are monstrous egoists . . . everyone works by himself here, without bothering others. Everyone wants to teach and no one wants to learn.'

He eventually gave up on Cauchy: 'Cauchy is mad and there is nothing that can be done about him, although, right now, he is the only one who knows how mathematics should be done.' He decided to cut his losses and head back to Norway, reaching the capital in May 1827. His great breakthrough had been completely ignored. He still had no position and was now heavily in debt. The prospect of his marrying his fiancée was looking even bleaker than when he had set off on his grand tour.

But Abel was still thinking about mathematics. He was beginning to grasp that there was much more to his solution of the problem of the quintic than revealing that there was no formula to solve these equations. How the roots of each equation behaved as you permuted them seemed to suggest that each equation had a certain symmetrical object associated with it. And it was the individual properties of these symmetrical objects that held the key to how to solve each individual equation. Abel was beginning to see the Strait of Magellan opening up before him, just as Degen in Copenhagen had predicted. He wrote to Crelle with details of his ideas, together with a request for a loan declaring that he was 'poor as a church mouse' and ended with the salutation 'Yours destroyed'.

Desperate to spend some time with his fiancée, Abel decided to take a break from the pressures of mathematics, penury and uncertainty over his future. In December 1828 he travelled out to the island of Froland to spend Christmas with his fiancée, where she was a governess to a family. But he could not afford the warm clothes that the harsh winters on Froland demanded. After a romantic sleigh ride through the frozen landscape with only socks for gloves, Abel fell dreadfully ill.

Despite Cauchy's lack of interest in Abel's work, others in Paris had begun to wake up to his fantastic achievements. After learning that he was living in poverty in Norway, they wrote to the King of Sweden in a desperate attempt to secure a position for this young mathematician with 'so rare and early-developed talent'. Crelle too was fighting hard to secure Abel a position in Berlin. Finally, on 8 April, he wrote to Abel to tell him the good news that the university there would offer him a professorship in recognition of his pioneering work. 'You can be completely at peace with regard to your future. You are coming to a good country, a better climate, closer to science and to genuine friends who regard you highly and are very fond of you.'

The letter arrived too late. On the very day that Crelle sat down to inform Abel of the good news, Abel died at the age of 26. The secrets of symmetry lay just over the horizon, just out of view. He never realized his dream of marrying his fiancée. On his deathbed he wrote to his friend Baltazaar Keilhau, pleading for him to marry Christine Kemp in his place. Despite never having met her, his friend agreed.

As time passed, mathematicians began to take stock of the beauty and depth of Abel's work. As the French mathematician Charles Hermite commented, 'He has left mathematics something to keep them busy for five hundred years.' Abel was awarded posthumously the Grand Prix by the Paris Academy in 1830. Today, the greatest accolade for a mathematician is to be awarded the Abel Prize by the Norwegian Academy. Set up in 2003, the prize is worth half a million pounds to the winner and is meant to be as prestigious as the Nobel prizes are in the other sciences.

But for most mathematicians the prize is not getting the Grand Prix from the Paris Academy or a telephone call from the Norwegian Academy, but the rush of adrenaline that you get on making that elusive breakthrough. My student Christopher's prize came on September the 11th, 2001 when the palindromic symmetry appeared through the mist of his calculation. For Luke the trophy came with the beep of the computer heralding the first calculation of a zeta function without this symmetry. Abel may have died before being recognized with prizes and jobs, but he had the satisfaction of knowing that he'd cracked one of the great problems of mathematics.

February: Revolution

It was the best of times, it was the worst of times, it was the age of wisdom, it was the age of foolishness . . .

CHARLES DICKENS, *A Tale of Two Cities*

13 February, La Villette, Paris

As unlikely as it may sound, there are distinct mathematical styles. Despite the universal nature of the language of mathematics, different mathematicians use this language in different ways, and this reflects cultural traits. The Anglo-Saxon temperament tends towards the nitty-gritty, revelling in strange examples and anomalies. The French, in contrast, love grand abstract theories and are masters at inventing language to articulate new and difficult structures.

It was with the help of my French collaborator François Loeser that I was able to set the elliptic curve example, my gravestone epitaph constructed in Bonn, into a grand theory called motivic integration. With my discovery at the Max Planck Institute I dug the first tunnel connecting two lands, but the language I learnt in Paris from François has helped me to widen this tiny tunnel into a beautiful thoroughfare. So now the route between groups and geometry is as well served as the Channel Tunnel I've just travelled through to get to Paris.

Mathematics requires an obsessional personality. In addition to his amazing ability to trek over mathematical terrain, François does long-distance running – and I don't just mean marathons. He runs 48-hour races which require crossing from one side of an island to the other and climbing a 3,000-metre mountain in between. On one of my visits, he had just completed a 70 km run through knee-high snow. I went on a 'gentle run' with him once before breakfast and ended up throwing up in his garden on the last leg home. The determination he shows in his running is reflected in his mathematical stamina. He is able to sustain an argument through some of the most abstract terrain I've ever encountered.

François's other obsessions are things that I am happier to join in with – food, wine and Tintin – and even then I can't quite keep up with his depth of knowledge. 'Can you tell me who Belle is in Tintin?' he once challenged me. I tried to think of any woman in Tintin apart from Signora Castafiore, the opera singer, and could not think of any. Hergé, Tintin's creator, was something of a misogynist. 'It is the name of the horse that Tintin is riding on your T-shirt,' he said. I looked down at my T-shirt, which depicted a scene from *Tintin in America* with the young reporter dressed as a cowboy riding a horse.

On another occasion we went for lunch after a seminar I had given at the École Polytechnique. There was a beautiful round cheese on the table which I hacked into with great pleasure. An unnerving silence descended. François explained that you were only meant to cut this cheese in a horizontal plane, something which was almost impossible to do, which is probably why the cheese had remained untouched. And I thought Oxford food rituals were severe!

Some French mathematicians believe that the quality of their mathematics is a product of the French language in which it was written. One of François's colleagues, Bruno Poizat, is particularly proud of the French language and never bows to pressures from journals to write in English, the universally accepted language of science. One of his most important contributions is a seminal book on mathematical logic and its interactions with the theory of groups. His insistence on publishing it in French meant that no publisher would touch it. So he went ahead and financed the publication of the book himself under his own publishing name: Nur al-Mantiq wal-Ma'rifah, Arabic for 'Light of Logic and Knowledge'. Because he had complete editorial control the book is rather idiosyncratic. Every chapter starts with a pornographic picture. Poizat explains in the introduction that these pictures are there to soothe the brain before the difficult mathematics that follows.

You can imagine the outrage this caused. Mathematical logic has a large female contingent, and they were not happy, not least because when you look up some of the names of the women in the subject in the index, you get directed to the pages of pornography. The last chapter has a picture of the author in a dressing gown, leering out of an armchair at the reader. But the mathematics is so good that the book could not be ignored. In Poizat's view, the material is particularly suited to the language in which it is written:

> Scientific French, what a beautiful language! ... I have no French nationalist feelings, nor a nostalgia for the time when French had a more dominant position ... I believe that the plurality of languages in use for communication of science has a value per se.

At a conference I attended in Russia, Poizat insisted on speaking in French with simultaneous translation into Russian, and was obviously delighted to leave the English-only members of the audience in the dark:

> Well intentioned people have told me that it is quite rude to address a person he or she cannot understand. If this were true, the community of mathematicians would rate highly in the scale of rudeness considering the number of times some of its members have spoken to me in English.

This afternoon I'm hoping to drop in on François at the École Normale Supérieure to see whether his French perspective might help me make sense of when my zeta functions have palindromic symmetry and when they don't. We also have been collaborating on a paper that we have done the thinking for but neither of us has had the time to write up. But my primary reason for a day trip to Paris is not to explore mathematical structures. The French capital is blessed with some fantastic examples of symmetry in architecture, and I've taken some time out of my research to do a little pilgrimage across the Channel with Tomer in tow, my faithful Passepartout. Our first destination is a pyramid.

The extraordinary new entrance to the Louvre in Paris is a glass pyramid provocatively set against the ornate seventeenth-century facade of the original building – an inspirational juxtaposition (Figure

48). It is as if the visitor is being invited to emulate the great archaeologists, to be an Indiana Jones and plunder the great riches that the
Louvre has hidden in its depths. The structure of the Ancient Egyptian
pyramids is quite straightforward. Each layer provides a solid foundation for the next layer. But the pyramid at the Louvre is hollow inside.
Engineers have exploited the strength of the triangle to construct their
pyramid. Each triangular face is a lattice of smaller triangles and dia-
monds. There are over 600 panes of glass used in the Louvre's pyramid.
The pyramid is surrounded by water so that with the reflection one
sees not just a square-based pyramid but the octahedron, one of Plato's
symmetrical shapes.

After visiting the pyramid in the centre of Paris, we head up to the

Fig. 48 The pyramid at the Louvre, Paris.

Parc de la Villette, which is home to another extraordinary piece of architectural design. Surrounded by all the square boxes that make up most of suburban Paris, La Géode is a huge silver globe (Figure 49). Tomer is impressed: 'It looks like an alien spaceship.' Like the Alhambra and the Louvre, the use of water enhances the symmetry. La Géode floats like a huge bubble on top of an expanse of water that surrounds it. Inside, the sphere houses a huge Imax cinema which you enter from below ground level.

Fig. 49 La Géode at La Villette, Paris.

Nature is fond of the sphere as it is a shape with low energy – this is why bubbles and raindrops are spherical. But for humans, creating a sphere is not so easy. When Pope Benedict asked Giotto for a drawing to prove his worth as an artist, Giotto drew a perfect circle – freehand. For an architect, building a sphere is perhaps the ultimate challenge. Anyone who has tried to gift-wrap a football will have experienced some of the difficulties that face the architect.

From a distance, La Géode looks like a perfect sphere, but as you draw closer you see how the architect has achieved this impression. Its surface is built from triangular pieces. There are a total of 6,433 triangles. Different sorts of triangle have had to be used to cover the curved surface – 136 varieties in all. Some meet in hexagons while

others create pentagonal shapes. The polyhedral framework on which
La Géode is based is the icosahedron, the Platonic solid that is made
from 20 equilateral triangles. By subdividing these triangles into ever-
smaller triangles and making the pieces bow outwards, the architect
has got closer to a sphere.

Tomer enjoys seeing himself elongated then flattened in the bulging
glass. The whole thing is like a house of mirrors. An eerie, bell-like
music circles round La Géode, adding to the surreal effect. Apparently
it comes from a musical clock, and you can tell the time according to
the location of the sound of the bells in relation to the building.

When you are using triangles to create a complete sphere, it is
essential that there are no faults in the specifications of each three-sided
face. It was an extremely tense moment as the creator of La Géode,
dressed in a futuristic spacesuit, manoeuvred the last triangle into place
on 16 April 1984. One small error, and the spherical jigsaw would not
have fitted together. Mathematics and engineering combined to ensure
that La Géode had no embarrassing hole in its surface.

Liberté, égalité, fraternité

This French passion for building a sphere, the most symmetrical of
shapes, in the heart of Paris is not a recent phenomenon. La Géode
realizes a dream that goes back two hundred years. Paris in the first
few decades of the nineteenth century was a city at the centre of
dynamic changes sweeping through Europe. The revolution of 1789
had made the impossible suddenly possible. The *ancien régime* and
all its outmoded ideas had been swept away by the radical ideas of
revolutionary young minds. *Égalité* – equality – was the word everyone
was shouting from the barricades. All parts of society should be treated
equally. Symmetry at the very heart of society.

Architecture was regarded as an ideal vehicle for this new ethos. In
1784 Étienne-Louis Boullée had drawn up plans for the construction
of a huge sphere in Paris dedicated to Isaac Newton. The French
revolutionaries were particularly attracted by the egalitarian nature of
the sphere as a perfect symbol to embody their ideals and adopted his
plans. For them, the sphere was the ultimate socialist shape, for no
direction is favoured over any other.

Boullée's proposal was for a hollow sphere 150 metres across, the surface peppered with small holes to allow in light during the day (Figure 50). The aim was to create a picture of the night sky inside the sphere – the first planetarium. At night the inside would be lit by a huge lamp suspended in the middle of the sphere, like the sun that sat at the centre of the eighteenth-century universe. But the French revolutionaries discovered that, like politics, realizing the ideals of mathematics is not always so easy, and for two centuries creating the sphere remained a Parisian dream.

Fig. 50 A sketch of Étienne-Louis Boullée's sphere.

The fervour of that initial turbulent period eventually ran out of steam. Nevertheless, the revolutionary spirit was the inspiration behind many institutions founded at the time, many of which survive to this day. New centres of learning were set up to cultivate the new thinkers of this revolutionary age. The year 1794 saw the creation of the great École Polytechnique, a college that would transform the scientific education of the country and put Paris at the centre of the intellectual map of Europe.

Students at the new École would be chosen on grounds of intelligence and knowledge alone. No one would be denied a place for lack of funds. Every new student received a salary of 900 francs a year (about £1,000 today) together with travelling expenses equal to those received by a first-class gunner in the army. The intellectual army was as important to the new republic as its military force. The École even had its own military uniform. Of all the disciplines, mathematics and science were the most valued. Indeed, the motto of the École became *Pour la Patrie, les Sciences et la Gloire* – 'For Homeland, the Sciences and Glory'.

In 1804 Napoleon was crowned Emperor. The Revolution seemed to achieve some stability by tying its colours to the cult of Bonaparte. But Napoleon turned revolutionary politics into a mission to conquer the world. France expanded, and just as quickly contracted. Napoleon's humiliating defeat in 1814 put France back under the white Bourbon flag and the rule of Louis XVIII, brother of the guillotined Louis XVI. Singing the Marseillaise was forbidden, and the tricolour no longer flew over the rooftops of Paris.

The energy and power that the Revolution had unleashed across Europe could not be so easily reined in by those who supported the Restoration. The École Polytechnique continued to be a thorn in the side of the establishment and remained a centre of Jacobinism and Liberalism. Despite being stripped of its military status and the head of the school being sacked, the Royalists still recognized the school's importance in training the best scientists the country could get. Although the École Polytechnique was turning out scientists for the modern age, the *ancien régime*'s Academy of Sciences, founded in 1666, still represented the pinnacle of academic achievement. The Academy was closed during the early zeal of the Revolution since it was regarded as an organ of the Royalist cause but it was soon restored to its place at the centre of France's intellectual life.

In the spring of 1829, a 17-year-old schoolboy by the name of Évariste Galois made his way through the grand courtyard of the Academy, overlooking the Seine. He had a package to deliver to Professor Cauchy, containing a manuscript that the young student knew would interest not only Cauchy but the rest of the mathematical community. It took some audacity on Galois's part to choose to present his ideas to Cauchy rather than any other academician. The professor was known to present only his own work during the weekly sessions of the Academy, and was not usually sympathetic to the ideas of others. He had completely ignored Abel's work and subsumed Ruffini's work into his own discoveries. But Galois was convinced that the mathematical breakthrough he had made over the previous few months would spark a mathematical revolution to rival the one Robespierre had ignited in Paris.

Cauchy was used to receiving manuscripts from across Europe from unknown authors who hoped to have their work discussed at the regular sessions of the Academy. He'd waded through the 512-page proof sent by the medical doctor Ruffini in Italy; he'd battled with the condensed six-page account by the Norwegian Abel. But he must have been rather intrigued by the address from which this particular package arrived: the Lycée Louis-le-Grand. Both Robespierre and Victor Hugo had attended this school, which was housed in an imposing but rather dilapidated building on the left bank of the Seine, not too far from the Academy. The bars on the windows made it look more like a prison, and the regime within the school did little to deter such an impression. Discipline was strict, punishment frequent. There were a dozen bare cells, continually occupied by pupils caught talking or fidgeting in class or even simply turning over too much in bed.

Galois had grown up in the small village of Bourg-la-Reine on the outskirts of Paris. Up until the time he was sent as a boarder to the Lycée at the tender age of 12, he was mollycoddled at home by his mother. It must have been a painful transition. He was extremely close to his father and missed him terribly. Suddenly he found himself in a world where he was punished regularly, subjected to a prison diet of dry bread and water, and bitten at night by the rats that infested the school. Galois worked hard to meet his father's high expectations. Learning languages particularly appealed to him, and for the first three years he impressed his teachers with his grasp of Latin and Greek.

When he was 15, Galois came across a book written by the French mathematician Adrien-Marie Legendre which revealed a new language, one that seemed to speak directly to him. An exciting new world full of mystery opened up and offered a refuge from the horrors of the Lycée. The mathematics book was meant for a two-year course, but Galois devoured it in two days. His teachers certainly recognized a change in him: 'He is under the spell of the excitement of mathematics.' His humanities teachers reported of his other work that:

There is nothing in his work except strange fantasies and negligence; he always does what he should not do. He gets worse every day. He has

gone from one punishment to another. His ambition, his frequently
ostentatious originality, his bizarre character keep him isolated from his
school fellows.

Galois was certainly ambitious. His father, the mayor of Bourg-la
Reine, had been a great supporter of the Revolution and had instilled
in his son the ideals he himself held dear. At a time when the Bourbons
were in the ascendancy he was one of the few Liberal mayors elected
in France, helped partly by the indiscretions of the Royalist candidate,
who had been forced to flee the town.

Galois believed that the great École Polytechnique was his destiny:
a centre full of young revolutionaries and the powerhouse of academic
achievement in Europe. He wanted to escape the tedious routine of the
Lycée. So against his teachers' advice and without telling his parents, he
took the strict entrance exam for the École Polytechnique in June 1828,
aged 16. Unfortunately, ambition and arrogance were not enough. He
failed the exam.

There is a certainty about mathematics which rubs off on those who
can master its language. Galois had no doubts about his abilities, for
he could solve all the problems his teacher set him. There was no
ambiguity or room for discussion in the world of mathematics. His
proofs were right and he knew it. But mathematics is also about
communication. Everything might have been clear and transparent in
Galois's mind, but he also needed to convey his ideas to those around
him.

The rejection by the École Polytechnique spurred Galois on to prove
himself the following year. Candidates were allowed only two attempts
at the entrance examination, but next time he would show them that
he deserved his place at the most prestigious academic institution in
Europe. It was during this year that Galois encountered a paper that
would sow the seed for the breakthroughs contained in the package
he later delivered to Cauchy. It was the paper that had inspired Ruffini's
work on solving quintic equations, in which Lagrange started to
explore what happened if you permuted the five solutions around, like
shuffling a pack of cards.

Inspired by Lagrange's paper, Galois made the conceptual breakthrough that would take him and mathematics through the Strait of
Magellan and into a new world where symmetry would yield up its

secrets. Contained in the package he delivered to Cauchy at the Academy were his first attempts to articulate his ideas. This, Galois believed, would prove that he deserved a place at the École Polytechnique.

What shape is your equation?

Galois realized that there was a subtler question underlying the attempt to solve the quintic. Tartaglia had found ways to twist the cubic until it looked just like an equation that involved taking cube roots and square roots. Although Abel had proved that in general this was impossible to do for all quintics, there were still some equations such as $x^5 = 3$ that could be solved using fifth roots. But what about something like $x^5 + 6x + 3 = 0$? There should, Galois thought, be some way to distinguish between, on the one hand, quintic equations that could be manipulated until they could be solved by taking roots, and on the other, quintic equations that it was impossible to solve.

Abel had shown only that there was no grand formula that would solve all quintic equations in one go. He too had begun to think about Galois's more subtle formulation of the problem, but his untimely death had deprived him of the chance to explore this further. Galois realized that it was the symmetries of the solutions of the equation that held the key to answering this problem.

The equation $x^2 = 2$ has two solutions: the square root of 2, equal to 1.414 . . . , and the negative of this number. In the same way, with the introduction of imaginary numbers, a cubic equation, one with x^3 in it, has three solutions and a quintic – one with x^5 – has five solutions. The higher the largest exponent of x, the more solutions you get. Galois considered the four solutions of a quartic equation. For example, two of the solutions of $x^4 = 2$ are real numbers, namely 1.189 21 . . . and its negative, $-1.189 21$ . . . The two other solutions are imaginary numbers, $(1.189 21 . . .)i$ and $-(1.189 21 . . .)i$. We can draw a picture depicting these solutions on the map of the complex numbers that Gauss created (Figure 51). Here A and C are the two real solutions, and B and D are the two imaginary solutions.

The numbers are transformed in this picture into the corners of a square, and it is the symmetries of the square that hold the secret of why this equation can be solved so easily. Galois realized that there

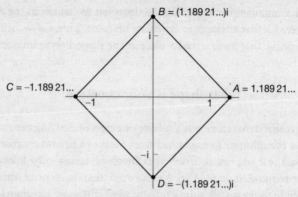

Fig. 51 The symmetries of the four solutions of the quartic $x^4 = 2$ correspond to the symmetries of a square.

were certain relationships between these four numbers that bound them together. For example, adding A and C gives zero, and so does adding B and D. So $A + C = 0$ and $B + D = 0$ can be regarded as 'laws' that relate the solutions of this particular equation. Each law produces a sort of rigidity in the picture.

Galois decided to look at the ways you can permute the solutions so that the new equation is still a law. For example, swapping A and C gives $C + A = 0$: that's still true. But look what happens if B and C are swapped. That turns the equation $A + C = 0$ into $A + B = 0$, but this new equation is false. This permutation isn't a law satisfied by the numbers A and B. There are potentially 24 different ways to permute A, B, C and D, but only eight of these permutations always change one law into another. For example, cycling the numbers is fine: if A goes to B, B goes to C, C goes to D and D goes to A, the laws $A + C = 0$ and $B + D = 0$ get changed into $B + D = 0$ and $C + A = 0$, and these are again laws which are satisfied by the numbers.

But these eight permutations of the letters aren't any old subset of the 24 possible permutations. They are in fact all the different permutations that describe symmetries of the square you get by joining the numbers A, B, C and D (Figure 51). The symmetries of the square are precisely the ways in which you can swap the numbers A, B, C and D and preserve the laws. The laws that the solutions obey are a little like the rigidity in the square – however the square is rotated or flipped, the corners A and C must end up opposite each other, as must corners B and D.

Galois did not have a clear vision of the possible shapes lurking behind an equation, or of why the language he was developing would help reveal the symmetry of those shapes. Perhaps it was just as well, because the power of the language lay in its ability to create an abstraction – a mathematical description that was independent of any underlying geometry. What Galois could see was that every equation would have its own collection of permutations of the solutions which would preserve the laws relating these solutions, and that analysing the collection of permutations together revealed the secrets of each equation. He called this collection 'the group' of permutations associated with the equation. Galois discovered that it was the particular way in which these permutations interacted with each other that indicated whether an equation could be solved or not.

When Galois took other quartic equations, such as $x^4 - 5x^3 - 2x^2 - 3x - 1 = 0$, he found that there were fewer laws connecting the solutions. Again, there are four numbers, A, B, C and D, which solve this equation. But this time there was less rigidity and it was possible to swap all the solutions around in any order and still preserve the laws. The 'group' of operations that Galois associated with this equation consisted of all 24 different arrangements of the four solutions. Again, there is a geometric object hiding behind this equation; this time it is the tetrahedron. The tetrahedron is the simplest of all the

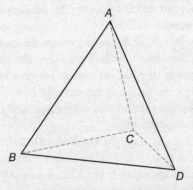

Fig. 52 The symmetries of the four solutions of the quartic $x^4 - 5x^3 - 2x^2 - 3x - 1 = 0$ correspond to the symmetries of the tetrahedron.

Platonic solids and has lots of symmetry. It is possible to put the four roots of the quartic at the four points of the tetrahedron (Figure 52). All the different symmetries of the tetrahedron correspond to permuting these four corners around. The symmetries of the tetrahedron and the permutations of the roots of the equations are actually two different manifestations of something abstract that captures the symmetry hiding behind both. And this is what Galois was beginning to articulate.

Galois's breakthrough was the discovery that the group of permutations associated with certain quintic equations had a particular character that made them rather different from quadratic, cubic and quartic equations. The shape hiding behind these quintic equations had symmetries that were much more complex than the symmetries of the tetrahedron or the square, objects that Galois found hiding behind certain quartic equations. For example, the symmetries of the five solutions of $x^5 + 6x + 3 = 0$ were closely related to the symmetries of one of the more complicated Platonic solids, the icosahedron. Galois had found that its symmetries were of a different order of complexity to those of the square and tetrahedron. The parcel he delivered to the Academy, addressed to Cauchy, contained his exposition of why things started to go dramatically wrong with equations involving fifth powers.

Missing manuscripts

When Cauchy opened Galois's package, his heart must have sunk at the sight of yet another bunch of papers claiming to prove that the quintic could not be solved. But as he perused the manuscript, he was rather taken by the ideas it contained, especially Galois's work on permutations. During the year 1812, while he was recuperating at his family's home in Paris following his collapse in Cherbourg, Cauchy had written two papers setting out a language and notation for the mathematics of permutations.

For example, Cauchy used the notation $(ABCD)$ to represent the cycling round of A to B, B to C, C to D and D to A. Similarly, (AC) would mean 'just swap A and C but keep B and D fixed where they are'. What Cauchy began to see was that there was a new arithmetic underlying these ideas. After all, doing $(ABCD)$ followed by (AC)

produced a third permutation: A goes to B, B goes to A, C goes to D and D goes to C – or in Cauchy's language, $(AB)(CD)$. He saw this as a sort of new multiplication which could be written as

$$(ABCD) \ast (CD) = (AB)(CD)$$

Cauchy had begun to explore the theory of this new language in a paper published in 1815. But the rather abstract nature of the paper had fallen on deaf ears. Cauchy was getting a reputation at the École Polytechnique for pushing students into abstract territory. The director had criticized his obsession with pure mathematics at the expense of teaching mathematics that would contribute to building Revolutionary France: 'It is the opinion of many people that instruction in pure mathematics is being carried too far at the École and that such an uncalled for extravagance is prejudicial to the other branches.'

As Cauchy read through Galois's paper, his initial scepticism gave way to real excitement at the mathematics it contained. Members of the Academy were certainly surprised when Cauchy stood up at the meeting on 25 May to register his intent at a future meeting to give a full account of Galois's ideas. The fact that Cauchy was willing to present anything but his own work is a testament to how important he must have thought it was, especially given that this was the work of a 17-year-old boy still at school. The academicians agreed that since Cauchy was best placed to analyse the work, he should take the only copy that existed home with him so that he could prepare his report.

A few weeks after the delivery of his manuscript, Galois would have his second chance to take the entrance exam of the École Polytechnique. Then he could realize his dream of joining the band of revolutionary students that he believed were the future of France. He could see the political camps in France squaring up to each other. The Royalists or 'Ultras' were gaining more and more political power, and Galois knew that the young students of the École Polytechnique would be on the front line, come the battle to stop the restoration of the *ancien régime*.

As Galois prepared for his examination, it was in fact his own family in Bourg-la-Reine that found itself on the front line. At the beginning of 1829, a young Catholic priest had arrived in the town where Galois's father, Nicolas, was mayor. Emboldened by the political mood in Paris, the young priest had banded together with the local Ultras to hatch a plot to bring down the Liberal Jesuit mayor. Nicolas Galois had a reputation for penning lines of verse to entertain his friends. So the priest started circulating forged vulgar verses attributed to the mayor. Despite proclaiming his innocence, Nicolas could not throw off the scandal and was eventually forced to flee the town. He rented a room just a few streets away from his son's school. On 2 July he committed suicide by hanging himself. A bomb had exploded in the boy's life, shattering the secure foundations he thought he was standing on.

Even in death, Nicolas Galois still commanded a lot of support in the town. Many of the townspeople went out to meet the coffin as it returned from Paris to Bourg-la-Reine and bore it to the church. Although he had committed suicide, the priest had agreed to conduct a service for the mayor and bury him in consecrated soil. Perhaps it was to assuage the guilt he felt for driving him to suicide. But others could not bear the hypocrisy of the priest who had orchestrated the mayor's death now ministering over the funeral. Galois witnessed the burial of his father descend into a political clash between the Royalists and Liberals, Catholics and Jesuits. Stones were thrown at the priest and insults were hurled as the coffin was laid to rest. Galois returned to Paris with the political fire bursting in his belly. But he had to try to focus on his impending entrance exam.

By all accounts, the examination before two professors of the École Polytechnique was a disaster. Galois showed no respect for what he regarded as two very mediocre mathematicians. Questions were met with a disdainful reply of 'that's obvious'. These things were probably obvious to Galois's sharp mathematical mind. What was not obvious to his political brain was that if he was going to be successful, he needed to play their game.

His head must have been in turmoil, having just buried his father in such traumatic circumstances. Perhaps the two examiners embodied for Galois the *ancien régime* he blamed for killing his father. One report of the examination even has Galois launching a board rubber across the room. Unsurprisingly, he failed to gain entrance to the Polytechnique for the second and final time. In the space of a month, the elation of his breakthroughs in the mathematical realm had been replaced by the breakdown of his hopes and dreams in the real world.

Instead of enrolling in the École Polytechnique to be trained for the political and academic elite, Galois had to make do with going to the École Préparatoire, a training college for schoolteachers. This was a reactionary and religious institution where failure to attend confession regularly resulted in expulsion. Now relegated to what Galois regarded as an academic backwater, his only hope was to wait for Cauchy's opinion of the manuscript he had deposited with the Academy.

Cauchy had arranged to deliver his report on the young mathematician's manuscript to the Academy on 18 January 1830. But the huge workload that Cauchy subjected himself to was beginning to affect his health. He missed the meeting and sent a letter apologizing for his absence:

> I was supposed to present today to the Academy first a report on the work of the young Galoi [sic], and second a memoir on the analytic determination of primitive roots in which I show how one can reduce this determination to the solution of numerical equations of which all roots are positive integers. Am indisposed at home. I regret not being able to attend today's session, and I would like you to schedule me for the following session for the two indicated subjects. Please accept my homage . . . A.-L. Cauchy

This is the last that was ever heard of Galois's first manuscript. The following week Cauchy only presented his own work. Galois never got his manuscript back, and it was never found among Cauchy's belongings. Niels Abel's treatise submitted to the Academy had also gone missing in Cauchy's possession, but eventually resurfaced after Abel's death.

When it comes to Cauchy's treatment of Galois, historians divide into two camps. One paints him as a self-centred and negligent man, interested only in his own work. The other suggests that Cauchy might have been the person who encouraged Galois to submit a revised manuscript for a new prize problem just announced by the Academy. The Grand Prix of the Academy was the highest accolade European science could bestow. The prize would be awarded to the most notable application of mathematics to general physics or astronomy or to an important analytical discovery. The closing date for submissions was the first of March. The board of judges included Siméon-Denis Poisson and Louis Poinsot, the man who had discovered the new symmetrical solid made from intersecting pentagons. We may never know whether it was Cauchy who encouraged Galois or whether Galois decided that, having waited months for Cauchy to reply, the only way to get a response out of the Academy was to submit a new manuscript for the prize.

Another member of the Academy, Jean-Baptiste Fourier, was appointed the new referee for Galois's second manuscript. However, Galois fared no better with his second judge. On 16 May 1830, a few weeks after taking custody of Galois's work, Fourier died. For the second time a manuscript of Galois's went missing, never to reappear. Without a manuscript or report Galois was never considered for the prize. He was never informed of the fate of his entry. In the end, this was the prize that Abel was posthumously awarded.

Revolution

Emboldened by the growing power of the Ultras, King Charles X decided on 26 July 1830 to dissolve parliament, rewrite the electoral laws to restrict voting to those with sufficient wealth, and suspend the freedom of the press. The ordinances were a red rag to revolutionary bulls who were already getting increasingly restless. The next day, four newspapers defiantly published articles denouncing the actions of the King and inciting rebellion. By the early afternoon the Parisian streets were full of crowds chanting abuse at the gendarmes sent to disperse them. Rocks starting raining down on the police, shots were fired and panic ensued. In the cross-fire a girl was shot dead. A worker picked up her limp body, placed it at the foot of a statute of Louis XIV and shouted for revenge. The city was soon gripped by a riot the like of which had not been seen in France since 1789. Barricades built from overturned carriages and furniture hauled from government offices once again blocked the streets of Paris.

As night descended on Paris, Galois, now a student of the École Préparatoire, could smell the fires burning on the barricades and hear snatches of the Marseillaise as revolution gripped the streets. The army was mobilized to contain the increasing number of citizens taking up arms and mounting the barricades. The rioters were joined the next morning by the revolutionary students of the École Polytechnique.

Although regiments of the army were guarding the entrances to the École, the students escaped by scaling the walls and took to the barricades, from where they led numerous bloody offences against the troops. The streets resounded to the songs of the young students: 'Fellow Frenchmen, let us sing of the heroic courage of the youth of the École Polytechnique.' By the afternoon the students had control of the Latin Quarter.

Here at last was the revolution that Galois had longed to be part of ever since the death of his father. But instead of joining his soulmates on the barricades, Galois was forced to sit and listen to the revolution from behind the closed doors of the École Préparatoire on the Rue Saint-Jacques, just streets away from the action. Although the lecturers at the École Polytechnique were only too happy to support the action of their students, the director of the École Préparatoire forbade any of his students to get involved. Galois and his classmates were locked in, imprisoned in their school and reminded of the promise they had made when they enrolled – a pledge of allegiance to the state. The director threatened to call in troops if necessary to keep his students from joining the insurrection.

Galois was incensed. By the evening of the second day of the revolution he could bear it no longer. It was too much to be listening to the students of the École Polytechnique making history while he was cooped up in the Préparatoire with a bunch of second-rate trainee schoolteachers, none of whom was willing to challenge the director's orders. That night, alone, Galois made an attempt to scale the walls but they proved too high. By the third day of what became known as the Three Glorious Days, the King's army had either deserted to join their fellow citizens at the barricades or fled alongside Charles X into exile. The white Bourbon flag was no longer flying over Paris. Instead, the Republicans were again in control. The morning air was filled with a cacophony of church bells signalling victory for the revolution.

But there was a problem. The uprising had been so successful that Paris had the chance once again to resurrect the Republic of the 1789 Revolution. But for most moderate Republicans this was a risk they were not prepared to take. France had been isolated and brought to its knees by the rest of Europe for its previous excursion into Republicanism. Now was not the right time for another full-blown Republic: that would have to wait for the revolutionary tide that swept across the whole of mainland Europe in 1848. Instead, the leaders of the Three Glorious Days invited the Duc d'Orléans to become the new King, a King they believed would uphold the institution of government without attempting, like his predecessor, to assume too much power. Crowned Louis-Philippe I on 9 August, the King and the tricolour flag were presented together in a Liberal fudge.

For the hardcore revolutionaries, the restoration of the monarchy betrayed the sacrifice of nearly two thousand citizens who had died to bring down the Bourbon flag flying over Paris. Galois's experience of being trapped inside his college during the revolution pushed him further into the extremist camp. During the summer vacation at his family home in Bourg-la-Reine, Galois harangued his mother and siblings with fiery revolutionary speeches. The July revolution had failed, he declared. It was necessary to have another uprising, one in which he was determined this time to play a central role: 'If I were only sure that a body would be enough to incite the people to revolt, I would offer mine.'

Galois's increasingly revolutionary zeal came to a head on his return to Paris in the autumn of 1830 for the new academic year. In a letter to the newspapers he accused the director of the École Préparatoire of being a traitor to the Republic for his decision not to allow his students to mount the barricades. The director was swift to respond, writing to the Minister of Education: 'I have expelled Évariste Galois. In my concern for his undoubted talent for mathematics I tolerated his unconventional behaviour, his laziness and his very difficult character.' But he would tolerate it no longer. Galois was a free agent.

During his time at the École Préparatoire, Galois had made only one real friend: Auguste Chevalier. Auguste was in the year above Galois, and his brother was a student at the École Polytechnique. The three would discuss politics at great length. The Chevaliers were followers of a utopian political movement called Saint-Simonianism. But its reluctance to resort to violence to further its political aims did not appeal to Galois and his increasingly aggressive Republican stance. Instead, Galois sought out the more militant Republican group known as the Société des Amis du Peuple. Thanks to the government-controlled press painting the Société as a dangerous band of militants, shopkeepers would pull down the shutters at the mere sight of a member of the society walking down the street.

Galois also enlisted with the anarchic National Guard. Established during the height of the 1789 Revolution, the Guard was a militia outside the French army. With its own banners, music and uniform, the Guard was more like the military wing of the Republican movement. At last, Galois was able to don a military uniform like his brothers-in-arms at the École Polytechnique. But several months after his coronation, Louis-Philippe outlawed both the National Guard and the Société des Amis du Peuple. The King recognized the threat that both represented, dissatisfied as they were with the failure of the Three Glorious Days to restore a true Republican government. Meetings of the Société now had to be held behind closed doors.

For Galois, expulsion from the École Préparatoire was a liberating experience – although it did have the downside that he no longer received the government grant he was entitled to as a student. Instead, he decided to present a series of weekly public lectures that would raise funds. Held in the back room of a friendly bookseller, the lectures would give him the chance to publicize the mathematical breakthroughs he had made and which had fallen on deaf ears at the Academy.

Galois placed a newspaper advert announcing the first meeting, to be held at 13.15 on Thursday 13 January 1831 in the Caillot bookshop off Rue de la Sorbonne. It drew an impressive audience of nearly forty people. The Chevalier brothers were keen to support their friend, and they were joined by several members of the Société des Amis du Peuple, who were perhaps expecting Galois to use the lectures to promote their revolutionary cause. If they were hoping for political revolution, they were disappointed. After several weeks the audience disappeared. Galois had tried to explain his new ideas for a revolutionary language to transform the study of equations, and ultimately the theory of symmetry. But his lectures were as impenetrable as the manuscripts he had sent to the Academy.

It is possible, though, that one member of the audience was the academician Poisson, one of the judges for the Grand Prix that Galois had competed for a year before. Shortly after the first lecture, Galois was approached by the great mathematician and invited to submit a third manuscript explaining his new mathematical vision. Galois wrote a new introduction, and once again made his way across the courtyard of the Academy to deposit his paper with the secretary. Poisson and

Sylvestre Lacroix were appointed during the following day's sessions of the Academy to report back on Galois's third attempt to convince the mathematical elite of his breakthrough.

On trial

This time, Cauchy was no longer around to act as assessor. Whereas Galois was disappointed with the current government's lack of revolutionary zeal, for Cauchy the new regime was far too extreme for his tastes. Shortly after taking office, the government had insisted that public servants, including professors at the École Polytechnique, swear an oath of allegiance to the new regime.

Cauchy would not betray his deeply held religious and political beliefs and refused to bow to the new government's demands. Fuelled by his childhood memories of the traumas of the Revolution of 1789, Cauchy fled Paris on 30 August 1830. From his exile, first in Switzerland and later in Italy, he saw himself stripped of all the positions he had held in the Academy and the École Polytechnique. Cauchy was simply too afraid to return to Paris in the present climate, despite the fact that he had left his wife and children behind. He remained in exile for eight years.

With his manuscript now in new hands, Galois was once again hopeful that he might at last get recognition for his work. He even began attending the weekly meetings of the Academy in the hope of hearing a report on his work. His arrogant self-belief in his mathematical abilities emboldened him to speak up and comment on the work of other mathematicians, despite being only 19. He was getting quite a reputation for his aggressively critical interjections. After one particularly fiery exchange between Galois and a lecturer at the Academy, Sophie Germain, one of the few women to attend meetings at the Academy, wrote to console the lecturer: 'he has kept up his capacity for being rude, a taste of which he gave you, after your best lecture at the Academy'.

Several months went by, and Galois still had not heard his manuscript discussed. Unable to contain his impatience any longer, he blasted off a letter to the Academy's president which barely hid the anger that was obviously seething below the surface:

> The research making up this memoir is part of a work I had submitted
> for the Grand Prix de Mathématiques last year . . . the prize committee
> decided that I could not have solved it, firstly because my name is
> Galois and also because I am a student. I was informed that my memoir
> had been lost. This should have been a lesson for me. Nevertheless I
> partially rewrote it, and submitted it to you on the advice of a fellow
> of the Academy.

He went on to demand that Lacroix and Poisson either own up to
having lost the latest manuscript or at least indicate whether they
intended to report on it to the Academy.

His despair at the failings of the establishment eventually bubbled
over with spectacular consequences. On 9 May, members of the Société
des Amis du Peuple invited two hundred fellow Republicans to a
banquet to celebrate the recent release from custody of several
members of their organization. A charge had been brought against 19
members for sporting the uniform of the National Guard after the
decree by the King that the Guard should disband. The not-guilty
verdict at their trial had elevated the 19 to the status of national heroes.

The party resounded to the popping of champagne corks and an
increasingly daring series of speeches and toasts to the Republican
movement. Galois, fired by an excessive amount of alcohol and the
charged political atmosphere, jumped to his feet, spurred on by his
young contemporaries. He raised his glass and shouted, 'To Louis-
Philippe!' Several guests began to jeer him for toasting the King they
wanted to depose. But then several others noticed a glint from Galois's
other hand. It was a small dagger. Galois's toast was an outright threat
on the King's life. The jeers turned to cheers, and the banqueting hall
erupted. Several of the less extreme Republicans realized that the party
was turning dangerous, and leapt from the restaurant's windows to
escape before the troops arrived.

The next day, Galois was arrested for incitement and threatening
the life of the King. He was brought to trial on 15 June but managed
to escape a prison sentence thanks to the quick thinking of his defence
lawyer. Despite Galois's wish to be sacrificed for the Republican cause
and be sent down as a martyr, his lawyer protested that Galois had in
fact qualified his threat. In the disturbance that followed his initial cry
of 'to Louis-Philippe', the guests had missed him add 'if he turns

traitor'. His lawyer argued that Galois was just considering a hypothetical situation and hadn't intended to make any threat against the King's
life. But this had been lost in the commotion that had erupted following Galois's first words.

The incident did end up having mathematical repercussions. To
support their friend at his trial, the Chevalier brothers published an
article in *Le Globe*, a newspaper sympathetic to the Saint-Simonian
movement. In it they deplored any recourse to violence, and threats
to the King's life, but offered various mitigating circumstances: Galois
was a mathematical genius whose work had been ignored or lost by
the establishment. 'He felt the germs of a brilliant future but with
neither protectors nor friends, he nurtured violent hatred of the
regime,' they wrote. The article documented the many times Galois
had submitted his ideas for comment at the Academy but had received
no response. Even now, the manuscript was with 'M. Poisson who is
to examine it but the wretched author has been waiting for a kind
word from the Academy for more than five months'.

Seeing his name in the newspaper must have goaded Poisson into
action, but the tone of the article didn't lead him to look upon Galois's
manuscript favourably. Instead, the report he delivered to the Academy
declared the 'thesis neither clear enough, nor sufficiently developed to
enable us to judge its rigour. Neither are we able to provide a clear
idea of this work.' For the first time, however, Galois actually had his
manuscript returned to him.

A week later, on the eve of Bastille Day, Galois was arrested again,
this time for wearing the banned uniform of the National Guard and
carrying weapons, and was locked up for the night. His cell-mate made
matters worse for Galois by graffitiing the walls with political cartoons
and slogans against the King. This time the courts were not so lenient.
After three months awaiting trial, Galois was found guilty and sen-
tenced to nine months in the Sainte-Pélagie prison on the southern
edge of Paris.

Mathematical escapism

Literary escapism has long been a way for inmates to cope with being deprived of their physical liberty. The seventeenth-century allegorical novel *The Pilgrim's Progress* was penned in Bedford prison, where John Bunyan was incarcerated for 12 years. The Marquis de Sade, Oscar Wilde and Adolf Hitler all penned significant works while inside. But history also bears witness to a slightly more unexpected way to break the tedium of hours locked up with only one's brain for company: mathematical escapism. Several of the hostages incarcerated for years in the Lebanon in the 1980s described how exploring numbers in their heads helped relieve days of isolation.

In 1940, the pacifist and mathematician André Weil, brother of the French philosopher Simone Weil, found himself in prison awaiting trial for desertion. During those months in Rouen prison, Weil produced one of the greatest discoveries of the twentieth century, on solving elliptic curves. He wrote to his wife: 'My mathematics work is proceeding beyond my wildest hopes, and I am even a bit worried – if it is only in prison that I work so well, will I have to arrange to spend two or three months locked up every year?' On hearing of his breakthrough, fellow mathematician Henri Cartan wrote back to Weil: 'We're not all lucky enough to sit and work undisturbed like you . . .'

Finding himself locked up, Galois too sought escapism in his mathematics. During his incarceration he received the report by Poisson on his manuscript. Although quite negative, it did at least end with an encouraging final paragraph:

> The author claims that the proposition which is the subject of his memoir is part of a general theory rich in application. Often, different parts of a theory are mutually clarifying, and it is easier to understand them together than in isolation. One should rather wait for the author to publish his work in its entirety before forming a definite opinion.

Galois decided to rewrite the manuscript he'd got back, adding a new extended introduction. As he worked away on his mathematics he soon got a reputation amongst his fellow inmates for being the young scholar in their ranks.

But he couldn't keep to himself his frustrations at his shoddy treatment by the Academy of Sciences and ultimately by the Institut de France, which oversaw all the academies in France. His new text began to fill with words full of vitriol and anger:

> It is the men of science who are responsible for my manuscripts being lost in the files of the Institut de France. I fail to understand such negligence by men who have Abel's death on their conscience; not that I wish to be compared with that eminent mathematician.

But Galois didn't spend all his time railing against the mathematical establishment. He also wrote about the power of a new conceptual approach to mathematics, of a move away from the complicated calculations of Euler towards the 'elegance of modern mathematicians whose minds quickly grasp all at once a large number of operations'. His great breakthrough was to identify a new abstract entity, what Galois called a group. The true essence of an object's symmetries would be found not by focusing on them individually, but by studying them together as a group.

Galois battled with the problem that every mathematician has to face when writing up a new discovery. Put in too little detail, and readers will have not enough directions to help them through the new mathematical maze. Yet put in too much detail and you swamp the readers, who will then have no clear vision of where you are trying to take them. Abel's six pages were at one end of this spectrum, while Ruffini's 512-page epic lay at the other.

Galois realized that his account was rather short on explanation: 'the printer when he saw the manuscripts thought they were an introduction'. Yet he admits that it would have been

> all too easy to substitute all the letters of the alphabet in each equation which would have multiplied the number of equations indefinitely. After the Latin alphabet I could have used the Greek one and when this had been used up, we still have the German Gothic letters and nothing would stop us using Syraic or even Chinese lettering!

He was trying instead to communicate his understanding of the concepts: 'there is as much French as algebra', something that is true of many mathematical papers, and comes as a surprise to those expecting just a stream of equations. Galois, though, was still trying to find the mathematical voice that would bring his ideas alive in the minds of others.

While he was inside, Galois made friends with another member of the Société des Amis du Peuple, who also spent his hours in prison studying and writing. Seventeen years older than Galois, François-Vincent Raspail had already made a name for himself as one of France's leading natural scientists with an important classification of grasses and a new theory of the biological cell. In discussions with the young mathematician, Raspail became aware of the young man's talents: 'In two years time he will be Évariste Galois, the scientist! But the police do not want scientists of this calibre and temperament to exist.'

The other prisoners were not so respectful, and enjoyed teasing the inexperienced young revolutionary. Several times they challenged Galois to drinking contests. As Raspail described in letters from the prison:

> To refuse the challenge would be an act of cowardice. And our poor Bacchus had so much courage in his frail body that he would give his life for the hundredth part of the smallest good deed. He grasps the little glass like Socrates courageously taking the hemlock; he swallows it at one gulp, not without blinking and making a wry face. A second glass is not harder to empty than the first, and then the third. The beginner loses his equilibrium. Triumph! Homage to the Bacchus of the jail! You have intoxicated an ingenious soul, who holds wine in horror.

In another letter sent from the jail, Raspail describes Galois drunkenly opening his heart:

> How I like you, at this moment more than ever. You do not get drunk, you are serious and a friend of the poor. But what is happening to my body? I have two men inside me, and unfortunately, I can guess which is going to overcome the other. I am too impatient to get to the goal. The passions of my age are all imbued with impatience . . . See here! I do not like liquor. At a word I drink it, holding my nose, and get drunk. I do not like women and it seems to me that I could only love a Tarpeia

or a Graccha. And I tell you, I will die in a duel on the occasion of some coquette of low standing. Why? Because she will invite me to avenge her honour which another has compromised. Do you know what I lack, my friend? I confide it only to you: it is someone whom I can love and love only in spirit. I have lost my father and no one has ever replaced him, do you hear me . . . ?

The speech is so prophetic that one can only guess that Raspail's letter was edited in the years after Galois's death. But Galois almost didn't make it out of jail to meet his destiny. One evening when he was going to bed, a gunshot flashed across the prison yard and a man in his cell fell to the ground. It appeared that someone in the garret opposite, where the guards were stationed, was picking off prisoners. When the guards on duty eventually arrived, the prisoners were in uproar. Galois in particular was incensed, convinced that the shot was meant for him. He accused the warden of deliberately organizing the assassination of difficult prisoners. The warden, seeing that the prison was about to erupt into a riot, promptly had Galois thrown into a solitary cell in the very depths of the prison.

The other prisoners protested vociferously at Galois's treatment: 'You throw into the dungeon both the victim of this shameful trap and the witness of it? This young Galois doesn't raise his voice, as you well know; he remains as cold as his mathematics when he talks to you.' 'Galois in the dungeon!' cried another. 'Oh, the bastards! They have a grudge against our little scholar.' The evening ended in a full-scale riot.

Love in the time of cholera

The unworldly nature of mathematics very often rubs off on those who spend a long time in its realm. Galois's fear of women, as expressed to his friend Raspail, was probably a result of his complete inability to grasp the rules and logic of the complicated game of love. His one experiment in this field had disastrous consequences.

It was not the assassination or imprisonment of political activists that finally put paid to a full-scale revolution in Paris, but the dreadful effects of a cholera epidemic that broke out in the spring of 1832.

Anyone with money fled the city; those in the slums suffered terrible losses. The authorities decided to move the young and sick prisoners out of the Sainte-Pélagie prison to avoid the inmates being wiped out by an outbreak of the infection. Galois was among a group of prisoners who were moved on 16 March to a clinic in the Latin Quarter. A month later, Galois completed his prison sentence. Although a free man, he decided to continue living at the clinic.

Having been locked up for months in the company of men, Galois now came into contact with a young woman at the hospital. Stéphanie, the daughter of the doctor at the clinic, used to help her father on his rounds. She was particularly taken by Galois, but it seems that he was unable to navigate the relationship that built up between the two of them. The elation of falling in love was quickly replaced by the despair at having his advances rebuffed by the young woman.

In mid May, Stéphanie wrote Galois two letters trying to cool his advances. Galois tore them up in a fit of rage and threw them into the fire. Then, regretting his actions, he tried to reconstruct what Stéphanie had written. On the back of some of the mathematical papers found after Galois's death are snatches of the correspondence from Stéphanie written in his hand: 'Please, let us put an end to this. I do not have the spirit to keep up such a correspondence but I shall try to find enough to converse with you as I did before anything happened.' Galois was left desperate by the end of the affair and wrote on 25 May to his friend Chevalier:

> How can I remove the trace of such violent emotions as I have felt? How can I console myself when in one month I have exhausted the greatest source of happiness a man can have, when I have exhausted it without happiness, without hope, when I am certain it is drained for life?

What happened over the next few days remains something of a mystery. Galois, a free man, was again attending meetings of the outlawed Société des Amis du Peuple. At a meeting on 5 May at which Galois was present, the society had decided that an armed uprising was the only way they could overthrow the new regime. That we know of such a decision is thanks to Lucien de la Hodde, a police informant who had infiltrated the society and reported on the plans being hatched.

What is slightly ambiguous from de la Hodde's report is whether Galois actually planned what happened next. On 30 May, in the early morning mist, a peasant walking by a pond on his way to market discovered a young man on the ground, writhing in agony. He had been shot, and had a single bullet wound to the stomach. It was, most likely, a duelling wound. In nineteenth-century Europe, duels were a common way of resolving disputes over women, politics, insults – even geese. Local newspapers would often carry notices of forthcoming duels, and their terms.

Galois was taken to the Cochin hospital where he died a day later, refusing to take the last rites offered him by the hospital's priest. 'Don't cry,' he said to his brother who was with him during the last hours. 'I need all my courage to die at twenty.' In letters sent to Republican friends the night before he ventured out to meet his death, he wrote:

> I beg my patriotic friends not to reproach me for dying in any other way than for my country. I die the victim of a cruel coquette and her two dupes. It is over a miserable piece of slander that I end my life. Oh! Why die for something so little, so contemptible? . . . I would like to have given my life for the public good.

It seems likely that the 'coquette' was none other than Stéphanie. What led to the duel is unclear. Had Galois discovered that Stéphanie had had a lover all along while she was, in his eyes, toying with him? The duel itself took place just a few streets away from the clinic where Galois had met her.

Although the duel was over a woman, there has been some speculation that Galois may have engineered his death to create the spark to ignite a new revolution. When the leaders of the Société des Amis du Peuple heard of his death, a meeting was summoned. According to the police agent among their ranks, it was decided that Galois's funeral would be the perfect excuse for the violent revolt they had been planning.

The next morning, three thousand people attended the funeral in Montparnasse. But during the funeral orations, word spread of an even greater cause for revolt. General Lamarque, one of Napoleon's right-hand men, had died that morning. His funeral was likely to whip up an even greater revolutionary fervour than this relatively unknown malcontent. A decision was quickly made. The revolt was put on hold, and Galois's funeral hastily drawn to an end. His death has to be one of the most pointless and tragic events in the history of mathematics.

Although several of the letters he wrote the night before the duel set out the reasons for the dispute, he spent most of the night trying to flesh out the mathematical theory he had failed to interest anyone in. He chose his friend Chevalier as the person he thought best placed to communicate his ideas. Galois seemed so certain about his impending death that it was with increasing panic that he spent the night trying to make his discoveries clear. He tried to address a number of the points raised by Poisson in his report. But as dawn approached he had to cut his explanations short. At one point he frantically wrote, 'There are a few things left to be completed in this proof. I have not the time.'

His letter to Chevalier ended with this desperate plea:

> In my life I have often dared to advance propositions about which I was not sure. But all I have written down here has been clear in my head for over a year, and it would not be in my interest to leave myself open to the suspicion that I announce theorems of which I do not have complete proof. Make a public request of Jacobi or Gauss to give their opinions not as to the truth but as to the importance of these theorems. After that, I hope some men will find it profitable to sort out this mess. I embrace you with effusion. E. Galois.

A few hours later he was fatally shot. Sophie Germain summed up in a letter to a friend the ill wind that seemed to be blowing through the mathematical community in Paris:

> There is decidedly a kind of fate or spell hovering over everything that has to do with mathematics. Your own difficulties, Cauchy's problems, M Fourier's death, as well as that of the student Galois, who, for all his impertinence, suggested certain exciting developments and tendencies.

Later generations would confirm Germain's hunch about Galois's ideas. Contained in the papers he left for Chevalier are the seeds of a totally new perspective on symmetry, one of the most fundamental concepts of nature. Looking through his notes now, I am amazed that such a young man could have had such a vision. Time and again, the big breakthroughs made by mathematicians in the last two hundred years in the theory of symmetry can be traced back to the profound ideas hiding in Galois's scribbled notes. This young revolutionary was the first to articulate a language that I now speak every day of my working life.

13 February, p.m., La Défense, Paris

Tomer and I drop by the École Normale, where François has his office. The École is at the heart of the Latin Quarter, Galois's home for the few years of his adult life. On the way to François's office you pass rooms with names above the doors of previous occupants: Samuel Beckett and Paul Celan. Tomer asks me who they were. I can do Beckett, but I'm rather stumped by Celan's claim to fame. François isn't in. I'll email him my questions instead. But seeing him isn't the main reason for our trip and it leaves us time to head up to the third stop on our Platonic Pilgrimage of Paris architecture.

We're off to see a cube. But this is no ordinary cube. This is a four-dimensional cube on the outskirts of Paris (Figure 53). The

Fig. 53 Arche de la Défense, Paris.

former President François Mitterrand was responsible for commissioning some of the great examples of modern architecture in Paris. For me, the Arche at La Défense is the most impressive and daring. La Défense was chosen as it lines up with some of the other great Parisian buildings along what is now called the 'Mitterrand perspective'. It starts with the pyramid at the Louvre, which we visited this morning, and proceeds via the Arc de Triomphe and the Egyptian needle to the huge Arche at La Défense.

As you climb the stairs at the métro exit, you see the Arche towering above the concourse at La Défense. It is so immense that the towers of Notre Dame could fit beneath it. Covering the side of a high-rise that borders the concourse is a huge advert depicting Thierry Henry smashing a football in mid air. It looks as though the Arche is the goal he's shooting at. But for me the significance of the Arche is not its size, but that it attempts to show what a cube looks like in four dimensions. Since we live in a three-dimensional world, it is impossible for us to construct a four-dimensional cube. But mathematicians have found other intriguing ways to capture these elusive shapes.

As we walk towards the Arche, the sun casts our shadows onto the pavement. A shadow is a two-dimensional picture of our three-dimensional shape. As we move and turn, the shadows change. Some shadows, like our profiles, give quite a good indication of what our bodies look like in three dimensions. The Arche exploits the idea that the Renaissance painters used to create the illusion of seeing

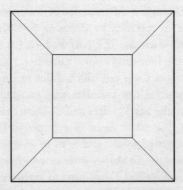

Fig. 54 A two-dimensional shadow of a three-dimensional cube.

three-dimensional shapes on a flat, two-dimensional canvas. If you want to depict a cube on a two-dimensional page, then a square drawn inside a larger square captures something of its three-dimensional shape (Figure 54).

The Arche takes this illusion one dimension up. A projection of the four-dimensional hypercube into three dimensions consists of a cube inside a larger cube. There is a strange effect as one walks towards these nested cubes. Although the weather is quite calm, there is a howling wind blowing across the square towards the Arche. Perhaps creating a shadow of a four-dimensional shape is rather dangerous. It feels as though the architect has opened up a tiny wormhole which is pulling us towards the centre of the Arche. Perhaps, instead of guiding our eyes towards the suburbs of Paris visible through the central cube, the Arche is in fact a portal to another world.

But there are other ways of describing this hypercube. As I'd discovered as a student, one way is by translating geometry into numbers. The coordinates of the four points of a square in two dimensions can be written as $(0,0)$, $(0,1)$, $(1,0)$ and $(1,1)$. The eight corners of a three-dimensional cube get translated via this language of coordinates into eight coordinates, $(0,0,0)$, $(0,0,1)$, $\ldots$, $(1,1,1)$. What this process of turning geometry into numbers can do is give a reality to something which in the visual language seems rather mysterious: the four-dimensional hypercube, a shape with 16 'corners', is given precise expression by the coordinates $(0,0,0,0)$, $(0,0,0,1)$, $\ldots$, $(1,1,1,1)$.

Although we can't see the geometry, the numbers allow us to explore the shape through a different set of mathematical lenses. An edge, for example, is specified by choosing two vertices that differ in one coordinate, such as $(0,1,1,0)$ and $(0,1,0,0)$. To find out how many edges the four-dimensional cube has, I simply count how many pairs of vertices there are which differ in one coordinate. So thanks to the algebra, I can calculate that in addition to these 16 vertices, the hypercube has 32 edges and 24 square faces, and is made up of eight cubes. But what about its symmetry? That was Galois's legacy. His new language would enable us to analyse the symmetries of these higher-dimensional shapes without ever holding them in our hands. It is this language that I shall need to exploit if I am ever going to see what symmetries in higher dimensions can be built from my simple prime-sided shapes. When Tomer falls asleep on the Eurostar back to London, I pull out my yellow pad, ready to make another onslaught on my problem.

March: Indivisible Shapes

> Mathematicians are like a sort of Frenchmen; if you talk to them,
> they translate it into their own language, and then it is immedi-
> ately something quite different.
>
> JOHANN WOLFGANG VON GOETHE

17 March, Stoke Newington

There are three stages to understanding something. The first is when
you suddenly get it. The second is standing in front of a seminar
audience and trying to convey to others the vision you've had. The
equations on the blackboard combine with the physical presence of
the speaker to conjure up ideas in the listener. But the third and
hardest stage of understanding something is translating it to the printed
page. There, the maths is going to be read without you present as a
guide. Everything must be well signposted so that the reader doesn't
get lost.

I'm currently in the thick of the third stage of understanding the
work I've been doing with Fritz. I'm trying to finish writing up some
of the insights we've had during our explorations in Bonn. It's only a
small step towards proving the grander conjecture I want to crack, but
it's a start. Just as Galois described, I must judge the amount of detail
so that the reader doesn't get so bogged down that they lose the
overarching narrative. This is the art of writing up your ideas, and it
often involves creating new language to describe them.

I was having great difficulty explaining to Dan, my DPhil supervisor,
a breakthrough I'd made in my early research. 'Why don't you give the

object a name?' he offered. It was a simple suggestion, but incredibly empowering. The act of naming the structure I was focusing on enabled me to express my ideas where before I'd been struggling. Thoughts suddenly seemed to crystallize. I could give a very precise definition of what this name meant which pinned it down without any ambiguity. The power of the mathematical language is that it allows me to capture a structure which was getting lost in the multiple strands of the logical argument.

The symbols that filled the books I looked at as a child are shorthand for ideas that can be articulated in longhand. But although this short-hand is very powerful, it is also what makes so much of mathematics impenetrable. It is like a Tower of Babel where each new storey introduces a new language. And if you skip one of the storeys in your ascent of the mathematical tower, you get more and more lost because you don't have a clue what people are talking about when they are using the words or notation introduced in the storey you missed.

Writing down the details of a proof calls for intense concentration. The proof should be like a piece of computer software. The brain is the hardware. The hardware can vary from person to person, but the program has to run on every machine. Something is wrong with a proof that keeps crashing when others begin to process it, so I've spent the morning building up the details of my logical argument. It's taken me weeks of hard slog to get this far, but I am nearing the end. It will soon be ready to be sent to a journal.

Galois was also aware that his ideas would be difficult for readers, and he wrote:

> This subject made necessary the use of new denominations, of new characters. We do not doubt that this snag will irritate the reader, from the very first passages. He will hardly forgive the author benefiting from all his trust for speaking a new language to him.

He was struggling to articulate what was in his mind in such a way that others could follow his reasoning. The one piece of writing that probably saved his ideas from disappearing into the mists was the passionate letter he wrote to Chevalier on the eve of his duel. Chevalier always had faith in his friend's mathematics and he could not ignore the plea from beyond the grave to seek the recognition he deserved.

Mathematical packages

Chevalier did not have the mathematical skills to sort out the mess
that Galois had bequeathed him, but he knew he had to try, both for
Galois's sake and probably for the good of the subject itself. Along
with the letter, there was a bunch of unfinished manuscripts that Galois
had left his friend.

Chevalier vowed not to let these discoveries fall into obscurity. With
the help of Galois's brother Alfred, who had been there at Galois's
death, they began to make copies of the most important passages in
order to send them to mathematicians across Europe. Galois had
asked Chevalier to seek the opinion of the two prominent German
mathematicians, Jacobi and Gauss. But Chevalier received no response
from Germany. Neither mathematician was prepared to spend time
trying to penetrate a dense unsolicited manuscript when they could
more usefully dedicate the time to developing their own ideas. But
Chevalier did not give up.

It took over ten years of trying before finally a French mathematician, Joseph Liouville, responded positively. Liouville was a professor
at the École Polytechnique, the institution that Galois had so desperately wanted to attend. An English mathematician who met Liouville
described him as 'a pleasant, chatty little man with whom I soon felt
at perfect ease. The only blemish I observed in him was an occasional
unmeaning giggle.' He was a prolific mathematician, publishing over
four hundred papers in subjects ranging from celestial mechanics to
number theory. He even has a number named after him:

$$0.110\,001\,000\,000\,000\,000\,000\,000\,001\,000\,0\ldots$$

where a 1 occurs at each decimal place which is one of the factorial
numbers $n!$. Liouville used this number to show that there are some
numbers which aren't solutions to equations: so-called transcendental
numbers. But as well as his own work, Liouville had established a
reputation for promoting the work of other, younger mathematicians.

Liouville had become rather disillusioned by the infighting that had
broken out within the French academies. He described how

a peculiar spirit of emigration has seized some critics and we have seen them heap abuse on one after the other of the men who in various fields of science have honoured France with great dignity . . . This sharp and peremptory style . . . will never be mine, for it dishonours both the character and talent of those who adopt it.

He decided to take things into his own hands, and founded his own journal in Paris to counter the wrangling that had infected the other French publications. Liouville had published many of his own papers in August Crelle's new journal in Berlin; this was the journal that had championed the work of the young Abel. Liouville hoped to do the same thing with his new *Journal de Mathématiques Pures et Appliquées*.

In 1842 Liouville got his chance to match Crelle's discovery of the work of the young Abel. Chevalier had sent him Galois's manuscripts, impressing on him how important he believed the contents to be. After looking through them, Liouville had a hunch that they contained something worth exploring and he decided to dedicate some time to untangling the mess of equations and arguments. His efforts were rewarded: 'Then in one moment I experienced the intense pleasure when, after having filled in all the careless gaps, I recognized the complete accuracy of the method proposed by this Galois and in particular the beauty of his theorem . . .' Suddenly, Galois's buggy proof had worked on someone else's mental machinery.

By 4 September 1843, Liouville was ready to present Galois's work to the Academy, just as Cauchy had promised to do over a decade earlier: 'I hope to interest the Academy in announcing that among the papers of Évariste Galois I have found a solution precise as it is profound of this beautiful problem: whether or not an equation can be solved by radicals.' Three years later, Liouville used his journal to publish Galois's ideas. With an interpreter at last to communicate this new language to the old guard, Galois's vision revealed a new world. Finally, mathematicians had found their way through the Strait of Magellan.

Even Carl Jacobi, one of the German mathematicians Galois had told Chevalier to contact, finally wrote to Galois's brother enquiring about the other work still unpublished, after he'd seen the papers in print in a mainstream journal. Alfred replied, thanking Jacobi profusely for his interest. He hoped that this would mean that a bit of his brother would live on beyond the grave through the recognition of his work. As G. H. Hardy once wrote, 'Archimedes will be remembered when Aeschylus is forgotten, because languages die and mathematical ideas do not. "Immortality" may be a silly word, but probably a mathematician has the best chance of whatever it may mean.' The fact that every mathematician now learns the fundamental concepts at the heart of Galois's work is testament to Hardy's belief.

Prime symmetry

What comes out of the subsequent development of Galois's work is a new way to think about symmetry. For most people, symmetry is a static property of an object. Galois's work inspired a different perspective. Symmetry should be thought of as something active rather than passive. A symmetry of an object is what you can do to an object to leave it essentially looking like it did before you touched it. This is something I started to understand as a kid. Take a seven-sided coin (such as the British 50p piece), place it on a piece of paper and draw an outline round it. A symmetry of the coin is any way of moving it so that it ends up back inside its outline. The magic trick moves.

At the heart of Galois's vision is the recognition that one shouldn't just look at individual symmetries of an object or a system. Rather, one must understand the symmetries in totality as a collection, or what he called a group. One symmetry should be thought of as something you 'do' to a structure, one of the magic trick moves. But Galois was interested in the collection of *all* symmetries and seeing what happens if you 'do' one symmetry after another. He discovered that it is the interactions between the symmetries in a group that encapsulate the essential qualities of the symmetry of an object.

These interactions led to a multiplication that binds all the symmetries together within the group. Galois saw that, just as two numbers can be multiplied together to give a third number, there was a way to combine two symmetries to produce a third symmetry. If I perform one symmetrical move followed by a second, the combined effect of the two moves is to create a third symmetry that I could have performed in a single move. For example, a clockwise rotation of a square through

90° followed by a horizontal reflection gives the same result as a reflection in a diagonal line (Figure 55). If we give these symmetries names, for example X for the rotation, Y for the horizontal reflection and Z for the diagonal reflection, we can express the relationship between these symmetries as a multiplication: $X * Y = Z$. In this way Galois discovered an extra layer of texture to the symmetries of each object reflected in how the different members interact with each other.

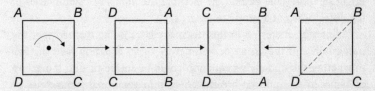

Fig. 55 A rotation followed by a horizontal reflection is the same as a reflection in a diagonal.

Galois wasn't interested in physical shapes. The corners of the square were replaced by solutions to equations. The letters A, B, C and D are then, for example, the four complex numbers that are the solutions to the equation $x^4 = 2$ (Figure 51, page 184). But the same principle applied. Instead of the rigidity of the shape, Galois was exploring the laws satisfied by the solutions, such as $A + C = 0$ and $B + D = 0$. He considered the group of permutations of the letters that preserved the laws $A + C = 0$ and $B + D = 0$ satisfied by these numbers. It was not the individual permutations that were important but rather their interactions. Take two permutations of the letters A, B, C and D which preserve the laws. For example the first permutation might send A to B, B to C, C to D and D to A; this permutation changes one law into another satisfied by the numbers represented by A, B, C and D. The second permutation might swap A with D and B with C, and that also preserves the laws. Then the combined effect of these permutations gives Galois a third permutation which fixes B and D but swaps A with C. The point is that this third permutation will also be in Galois's special group of permutations that preserve the laws.

Galois's amazing discovery, which he made when he was analysing the interactions of the symmetries within entire groups, was that some groups of symmetries could be broken down into smaller groups of symmetries. Other groups, however, seemed to be immune to such subdivision. For example, the rotations of a 15-sided coin can be built out of combinations of the rotations of a pentagon and a triangle, but the rotations of a 17-sided coin could not be broken down in any similar fashion. So the rotations of prime-sided shapes give prime indivisible symmetries. But Galois discovered other interesting indivisible symmetry groups beyond the simple prime-sided shapes. For example, the 60 rotations of a dodecahedron cannot be divided into smaller groups of symmetries. In particular, a group of symmetries might be indivisible despite the fact that the *number* of symmetries of the object was not a prime number.

Galois was interested in this discovery because he realized that the indivisibility of the group of symmetries of an equation held the key to whether the underlying equation could be solved or not. If a group of symmetries could be broken down into rotational symmetries of prime-sided shapes, the equation could be solved. Otherwise, it couldn't be solved.

For example, there are three complex numbers A, B and C which solve the cubic equation $x^3 + 2x + 1 = 0$. There are various laws satisfied by these three numbers: $A \times B \times C = -1$ is one of them. There are six different ways to permute these three numbers. The group of symmetries of the equation $x^3 + 2x + 1 = 0$ are those permutations that preserve the laws satisfied by the three solutions A, B and C. This particular equation, $x^3 + 2x + 1 = 0$, has the property that all six permutations preserve the laws connecting the solutions. The symmetry group of these three solutions is actually the same as the symmetry group of a triangle. Think of A, B and C as the three points of the

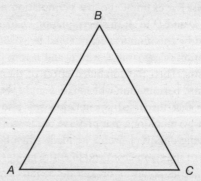

Fig. 56 The symmetries of the triangle give all six permutations of A, B and C.

triangle (Figure 56); every permutation of A, B and C can be obtained as a rotation or reflection of the triangle.

In Galois's language, the reason that Tartaglia was able to find a formula to solve the cubic is precisely because the six symmetries of the triangle can be 'divided' by the rotations of the triangle (which has three elements) to leave the reflections in a line. So the six symmetries of the triangle are built from two prime groups of symmetries, one with three symmetries and the other with two.

If we look at the more complicated case of the quartic equation solved by Cardano's student Ferrari, then the permutations of the four solutions A, B, C and D can be interpreted as the symmetries of a tetrahedron, where the solutions sit at the four vertices. Every permutation of the solutions A, B, C and D can be effected by a rotational or reflectional symmetry of the tetrahedron. There are 24 different symmetries in this group. Figure 57 shows the symmetry in which the tetrahedron is spun through 180° around an axis passing through the centre of two opposite edges. The effect of this symmetry is to swap A and B around and to swap C and D.

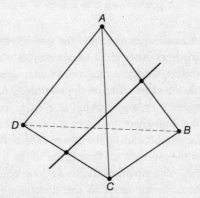

Fig. 57 The rotational symmetry of the tetrahedron that swaps A with B and C with D.

There are two other axes joining opposite edges whose symmetries swap the other ways of pairing these letters: one swaps A with C, and B with D; the other swaps A with D, and B with C. The intriguing thing is that this subgroup of symmetries is actually identical to the symmetries of a rectangle. A rectangle has four symmetries (Figure 58). They include, for example, a reflection in the horizontal line. This swaps A with D, and C with B. There are two other symmetries which swap the other ways of pairing these letters: a reflection in the vertical swaps A with B, and C with D; and a rotation of 180° swaps A with C, and B with D. The fourth symmetry is of course the symmetry that leaves the rectangle untouched.

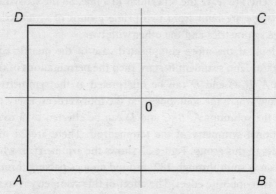

Fig. 58 The symmetry group of the tetrahedron can be divided by the symmetry group of the rectangle.

One would not have expected to find the symmetries of a rectangle hiding in a tetrahedron. After all, a tetrahedron is made up of four triangles. Nevertheless, the three rotations of a tetrahedron described above have exactly the same effect on the vertices A, B, C and D as if they were points on a rectangle. What is even more interesting is Galois's realization that there is a way to divide the group of 24 symmetries of the tetrahedron by the group of four symmetries of a rectangle, and get the group of six symmetries of the triangle that was hiding behind the cubic equation. Both the four symmetries of the rectangle and the six symmetries of the triangle are in turn built from symmetries of prime-sided shapes. This ability to reduce the symmetries to those of prime-sided shapes is the reason that the quartic equations can be solved by the formula that Ferrari discovered.

Dividing the tetrahedron by the rectangle to get a triangle is rather like the fact that 4 divides into 24 exactly, giving the answer 6 with no remainder. But dividing by groups of symmetries turns out to be much more subtle than dividing numbers. You have to be quite careful about how you break groups of symmetries down. For example, it seems clear that the symmetries of a triangle are hiding as a subset of the symmetries of the tetrahedron. But when you try to divide by the triangle, the operations you're left with don't make any sense as a collection of symmetries of an object. In contrast, there are lots of numbers that divide exactly into 24 to give another number, for example 3, 4 and 8. With symmetries, things are subtler. Only some subgroups of symmetries will divide to give another shape. So although there is a subgroup with three symmetries and one with four symmetries, and one with eight symmetries, only the one with four symmetries actually divides the group of symmetries of the tetrahedron exactly to leave another group of symmetries.

Galois's great breakthrough was his discovery that the shape hiding behind the quintic equation is the first example of something that cannot be divided into symmetries of prime-sided shapes, whatever choice of sub-objects you try to divide it by. For example, the equation $x^5 + 6x + 3 = 0$ has five solutions. There are not many laws connecting these solutions, and it turns out that the laws will be preserved no matter what way the solutions are reordered. There are in total $120 = 5 \times 4 \times 3 \times 2 \times 1$ different permutations of these five numbers, so the group of symmetries consists of 120 different operations.

It is possible to divide this group into two smaller groups, one with 60 symmetries and one with two symmetries. Just as numbers are either odd or even, permutations can be odd or even. It is possible to achieve any permutation by doing a sequence of swaps where two objects are interchanged at a time. For example, if I have a pack of five cards with the letters A, B, C, D and E on, and I want to cycle them to get them in the order B, C, D, E, A, I can do it by first swapping B with A, then C with where A is now, then D with A's new position, and E with A. With four swaps I've ended up with the same result as if I'd cycled the cards around:

$$(ABCDE) \rightarrow (BACDE) \rightarrow (BCADE) \rightarrow (BCDAE) \rightarrow (BCDEA)$$

Because it took four swaps, an even number, we call the permutation that cycles the five cards an even permutation. To cycle only four cards around, it takes three swaps, an odd number, so that permutation is called odd. Of the 120 permutations of the five cards, 60 are even and

60 are odd. Notice in particular that doing one even permutation followed by another gives me a third even permutation. This is not true for odd permutations: the combination of two odd permutations gives an even permutation. So it is only the set of even permutations that makes up a group of symmetries. It is in fact the same as the 60 rotational symmetries of the dodecahedron. One can match each of the rotations with an even permutation in such a way that the interactions between the different rotations are exactly mirrored by the interactions of the permutations. The symmetry of the shape can be used to do all the even shuffles of the pack of five cards.

What Galois discovered is the amazing fact that this symmetry group with 60 symmetries is indivisible. So although the number 60 can be divided by 5 to give 12, there is no subset of symmetries that divides into it to give a sensible set of operations that are the symmetries of another object. You might think that the rotations of one of the pentagonal faces should divide it. They certainly do form a subgroup of symmetries. But when you try to divide by this subgroup, the resulting division does not correspond to a group of symmetries. The 60 symmetries of the dodecahedron are so intricately related that it somehow creates this indivisibility. The symmetries are bound up in such a manner that the way two combine to give a third symmetry does not allow a strategy for breaking the group of symmetries down into sub-pieces.

So despite 60, the number of symmetries, being a very divisible number, the actual group of symmetries is as indivisible as if it were a prime number. And Galois recognized that this indivisibility meant that the underlying quintic equation couldn't be solved with a simple formula.

This was the great advance made by Galois, and what Abel had missed. The group of symmetries of an individual equation provided a way to tell whether that equation could be solved or not. If the building blocks that made up the group of permutations of the roots consisted of prime-sided shapes, then the equation could be solved by taking square roots, cube roots or higher roots of numbers. But if the building blocks involved some of these new indivisible shapes, such as the rotations of a dodecahedron, then there was no way to solve the equation with simple roots. This is why modern mathematicians give the name soluble groups to groups built from prime-sided shapes, to indicate that they are connected with equations that have solutions.

The group of even permutations of five letters became known as the alternating group of degree 5. It is the first building block in the periodic table of symmetry, after the prime-sided shapes. And it began a whole new way of looking at the world of symmetry.

Card tricks

Although Liouville was the first to spot the importance of what Galois had done, it was another French mathematician, Camille Jordan, who truly recognized the brilliance of his idea and started to build on Galois's foundations. Galois had discovered that every group of symmetries was either indivisible or could be broken down into smaller indivisible groups of symmetries. Mathematicians could now try to make a list of all the building blocks, like a periodic table of symmetry, and begin to address the question of how to use these building blocks to build new groups of symmetries. What was slightly unsettling was the realization that there were actually a whole variety of ways to put these building blocks together. This was completely in contrast to the way numbers worked. The building blocks of multiplication are the prime numbers, but with the primes 2, 3, 5 and 7, for example, the only number I can build is 210. But take prime-sided shapes with 2, 3, 5 and 7 sides, and there are in fact 12 different ways to put them together.

In 1870 Jordan published a book which crystallized the ideas Galois was alluding to. He also used the book to castigate the German mathematical establishment for failing to recognize the riches hidden in the texts they were sent by Chevalier. There may have been more than just mathematical rivalry at stake here. These criticisms were made just as the two countries were squaring up for the Franco-Prussian War, which broke out in the summer of 1870.

Jordan tried to see what groups of symmetries could be built from prime-sided building blocks. He soon discovered that, even though these blocks were simple, the variety of things you could build with them was so complex that it is still part of the mystery that my own research today is dedicated to unlocking. But he also established that

Galois hadn't just found one new indivisible group of symmetries. The indivisible group of rotations of a dodecahedron that Galois had unearthed at the heart of the quintic equation was just the beginning of a whole infinite family of such groups.

The obstacle to solving the quintic equation was the group of even permutations that shuffle the five solutions around. This, Galois discovered, was an indivisible group – or what Jordan called a simple group. The word 'simple' here was used by Jordan not to indicate that the objects were straightforward, but to indicate that the group is a basic building block, not a compound group of symmetries made up of smaller symmetries.

Think of the permutations of the five solutions as like shuffling a pack of five cards. Put each solution on a card. A permutation which changes the order of the solutions is like shuffling the cards. But there is nothing special about taking five cards as opposed to some other number. Take a pack with 52 cards, and consider all the different shuffles. Any shuffle of the pack rearranges the cards into a new order. The connection with symmetry is to think of the pack as analogous to an object with 52 faces. Every spin of the object brings the faces into a new orientation. Galois's argument for why the group consisting of all the even shuffles of five cards was indivisible applies equally well to any pack of cards.

Alongside the indivisible prime-sided shapes, Galois had added a new infinite family of building blocks of symmetry. Take a pack of 5, 6, 7 or more cards. The group of all the even shuffles is a group of symmetries which cannot be divided into smaller symmetries. The even shuffles of a pack of n cards is called the alternating group of degree n. But this wasn't the only new family of indivisible symmetry groups that Galois had discovered. In one of the memoirs he left for his friend Chevalier, Galois described another sort of building block of symmetry or simple group.

Galois's new building block had a more geometric flavour than the shuffle symmetries. But there is a twist, because the geometry is based on a new sort of arithmetic that Galois had just discovered. Usually we think of geometry as consisting of lines and points. As Descartes had revealed, the geometry that we can draw on a piece of paper can be translated into pairs of numbers. A point on the page can be changed into two numbers using the same principle that defines a location on a map. The two numbers tell you how far you should travel in the east–west direction and then in the north–south direction.

In conventional geometry the two coordinates can range through the infinite set of real numbers represented by the number line. But in this new geometry the coordinates are highly restricted. For each prime number p, the set of points in the geometry are given by coordinates (x, y), where x and y have to be whole numbers between 0 and $p - 1$. So there are only finitely many points in this new geometry. So, for example, the set of points for the prime 7 consists of the coordinates (x, y), where x and y are each any of the numbers from 0 to 6 inclusive. To compute with this geometry you have to use a new sort of arithmetic called clock arithmetic or modular arithmetic on the coordinates, which is different from ordinary arithmetic.

Think of a heptagon with the numbers 0 to 6 arranged on the seven points, as in Figure 59. To add 4 to 5, we move 4 steps round the heptagon from 5, which gets to 2. We write this as $4 + 5 = 2$ (modulo 7) to indicate that the heptagon was used to do the calculation. Subtraction follows a similar process. Multiplying 4 by 5 is the same as adding 4 to itself 5 times, so we move round the heptagon from 4 to 1 to 5 to 2 and get to the final answer of 6. But perhaps the most interesting thing about modular arithmetic is the possibility of doing division. In normal arithmetic, division necessitates producing a whole new set of numbers – the fractions. But because there is a prime number of elements in this set of numbers, you can always do division

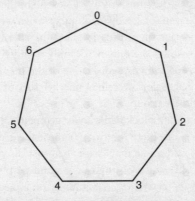

Fig. 59 Modular arithmetic. $4 + 5 = 2$ (modulo 7), $4 \times 5 = 6$ (modulo 7) and $3 \div 4 = 6$ (modulo 7).

with them. For example, 3 divided by 4 is 6, because 4×6 gets you to 3 on the heptagon (Figure 59).

What does a geometry based on these numbers look like? We can construct a 7×7 grid consisting of 49 points, which we can label by coordinates such as (1,2) and (4,4). Galois was interested in the lines that pass through the point (0,0). There are in fact only eight lines that can be defined in this geometry. To generate these lines, pick any point in the geometry and join it to the point (0,0). The other points on this line are found by continuing the line you've drawn. Because this is a finite geometry, you have to think of a line leaving the top of the grid and entering again at the bottom. Figure 60 shows one of the eight lines in the finite geometry got by joining (0,0) to the point (1,2). A line in this geometry consists of a subset of all the points in the geometry which can be connected in this way. Notice that if I took any other point on the line in Figure 60 as a starting point, despite the picture looking very different I would pick out the same seven points and therefore the same line in this geometry. There are therefore only eight lines in this geometry that pass through (0,0). Each line has six points on it in addition to (0,0). Since the geometry consists of 48

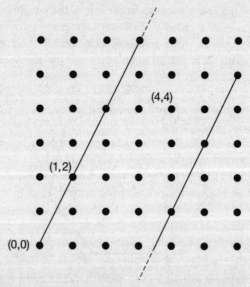

Fig. 60 A line consisting of seven points in Galois's finite geometry.

points in addition to $(0,0)$, and 48 divided by 6 is 8, that gives eight lines.

Galois began to investigate how the arithmetic of these numbers can be used to permute the eight lines in interesting ways. He had already discovered that you could generate interesting groups of symmetries from the shuffles of a finite collection of playing cards. So one option was to think of these eight lines as a set of eight cards and look at all their permutations. But that gives nothing new. The point was that using modular arithmetic with these numbers produced an interesting subset of all the permutations of the lines. Galois discovered that the group of symmetries of these lines produced by applying this new arithmetic was a new indivisible group.

Each symmetry is described by four numbers, which we can denote by a, b, c and d, arranged in a 2×2 grid called a matrix:

$$\begin{pmatrix} a & b \\ c & d \end{pmatrix}$$

The way this symmetry swaps all the points in the geometry round is to take a point, say (X, Y), and send it to the point $(aX + bY, cX + dY)$. For example, the matrix

$$\begin{pmatrix} 2 & 1 \\ 0 & 2 \end{pmatrix}$$

would send the point $(1, 2)$ to the point $(4, 4)$. If you choose the (a, b, c, d) in such a way that $ad - bc$ is not zero, then this interchanges all 49 points in the 7×7 grid. But the scrambling retains aspects of the geometry, because the seven points sitting on the line passing through $(1, 2)$ get sent to the seven points sitting on the line through $(4, 4)$. The matrix simply shuffles the eight lines through the point $(0, 0)$. Galois discovered that these symmetries could be used to construct a new building block of symmetry, or what Jordan called a new simple group. It had 168 symmetries but was indivisible, as though it were prime.

When Galois came across these indivisible groups of symmetries it was in the context of solving equations. Equations of degree 8 (i.e. starting with x^8) have eight solutions. To understand whether the equation could be solved, Galois needed to study the permutations preserving the laws satisfied by these eight solutions. While he was

exploring a special sort of equation of degree 8, he discovered that its
permutations were swapping the solutions around in just the same
manner as the eight lines were being permuted in his finite geometry.
The fact that Galois could prove that this special group of permutations
was indivisible meant that these special equations couldn't be solved.
That a 20-year-old could be so insightful and recognize that this would
produce such interesting mathematics is phenomenal. He saw that
whenever you took a prime number p, created a geometry with $p \times p$
points, looked at the $p + 1$ lines running through the geometry and then
took their symmetries, you got a new indivisible group of symmetries –
a simple group.

Interestingly, if you use the prime $p = 5$, you get a new view on
an old group. Although the group is permuting six lines, not every
permutation is possible. The group of symmetries you do get is simply
another way of constructing the group of symmetries of the even
shuffles of five cards or the rotations of a dodecahedron. Here is the
real strength of Galois's new language: three very different looking
symmetries can actually be recognized as three different manifestations
of the same underlying group of symmetries. It is a more abstract
example of how different designs in the Alhambra can still have the
same underlying symmetry. Although geometry based on the prime 5
gave Galois nothing new, as soon as he looked at 7 and higher primes
he found a whole new family of simple groups.

This new family of building blocks is known in the trade as $\mathrm{PSL}(2, p)$.
In his book dedicated to Galois's work, Jordan documented these new
groups and actually began to mine this new seam for more simple
groups. He saw new ways to twist these geometries so that they revealed
several other new families of simple groups. Although Liouville was
the first to highlight the brilliance of Galois's work, Jordan's treatise
was perhaps the most influential in establishing Galois's new vision.
This was an exciting time to be entering the world of symmetry. Two
young mathematicians who visited Jordan in Paris were particularly
taken by Galois's mix of geometry and symmetry.

The hardy and the handsome

Sophus Lie met Felix Klein when he came to Berlin on an academic visit from his home in Norway in 1869, 40 years after Abel's death. Like Abel before him, Lie had been given money to expand his mathematical horizons by undertaking a grand tour of the great academies of Europe. Lie had spent some time trying to sort out what to do with his life. At first he'd wanted a military career, but bad eyesight denied him a commission, so he enrolled at the university that Abel had attended. He found that he was very good at a whole range of subjects, but couldn't make up his mind what to concentrate on.

Mathematics was not a subject which ever came easily to Lie. But it was the rush of adrenaline that he experienced on his first mathematical discovery that sealed his mathematical future. Inspiration struck in the middle of the night as he was toiling away on a problem of geometry. He was so excited by the breakthrough that he rushed over to a friend's house and woke him up, shouting breathlessly 'I have found it!' He wrote later of that moment of revelation that 'As a young man, I had no idea that I was blessed with originality. Then, as a 26-year-old, I suddenly realized that I could create.'

Thinking that someone might steal his idea if he waited for a journal to publish his proof, Lie decided to print the result at his own expense. But the friend he'd woken in the night persuaded him that if he was going to impress the mathematical establishment, he had to get his idea into the mainstream journals. Lie's first paper was accepted by Crelle's journal in Berlin, the journal that had first published Abel's work. Its publication propelled Lie into the limelight, and secured for him the grant that made possible his European tour.

In Berlin he quickly became friends with Felix Klein, a mathematician who shared his particular slant on geometry. For both Lie and Klein, the fundamental objects in geometry were not the points but the lines. This is why Galois's ideas of permuting lines in his finite geometries resonated so strongly with them.

In contrast to Lie's hesitant inroads into mathematics, Klein believed that he was born to be a mathematician: the fact that the date of his birth (25/4/1849) consisted of squares of primes ($5^2/2^2/43^2$) was for Klein an auspicious beginning. He stormed down the mathematical racecourse and, unlike Lie, already had his doctorate by the age of 19.

Klein was a tall and handsome young man; Lie was a hardy outdoor type. But despite their physical differences they shared similar mathematical tastes. The mathematical community in Germany was not particularly keen on Klein's rather discursive style, believing it lacked the steely exactness that is so valued in mathematics. Klein disagreed. He believed that

> The presentation of mathematics in school should be psychological and not systematic. The teacher, so to speak, should be a diplomat. He must take account of the psychic processes in the boy in order to grip his interest, and he will succeed only if he presents things in a form intuitively comprehensible.

He took much the same approach to mathematics in the academic arena. Lie appreciated the visionary aspect of Klein's work, and on their trip to France they came across the mathematics that enabled them to articulate their emerging view of geometry.

Lie went to Paris as part of his grand tour, following very much in Abel's footsteps. He spent some time trying to improve his French. He found the theatre a good place to train his ear because he could buy the text for the play beforehand. He enjoyed wandering the streets of Paris, listening and looking at everything. He also attended lectures, which he found much easier to understand than the locals on the street: 'Mathematical lectures are not difficult to follow in a foreign language.'

When Klein arrived in Paris a short time later, the two men immersed themselves in the local mathematical scene. Both Lie and Klein commented in a report back to Germany on the feeling of self-satisfaction that seemed to have settled on Paris. It seemed to be resting on the laurels of its previous generation of groundbreaking mathematicians, and lacking the desire to take things further. It was during this period that the mathematical mantle passed from Paris to Germany. But Klein and Lie did appreciate the clarity with which the French wrote their papers. It was a clarity they felt was missing from the rather unreadable, telescopic accounts that German mathematicians produced. 'The intention of a mathematical work is reasonably to be understood, and not simply to engender admiration for the writer,' was their view. My impression of the majority of seminars that I attend is that unfortunately the German style has in general won through.

It was Jordan's style that the two visitors were most excited by. He was just finishing the book in which he set out the many ideas buried in Galois's manuscripts, writing with the kind of clarity that Klein and Lie admired. In the idea of a group of symmetries, Klein found the perfect language in which to express his ideas on geometry. Galois had been interested in the permutations of the lines of his finite geometries in connection with his work on solving equations. Klein, however, was interested in the geometry of the lines as fundamental objects. He saw that once the group behind this geometry was identified, it would provide a powerful way to talk about what 'geometry' really was. This perspective would ultimately illuminate the story of the walls in the Alhambra. The patterns of tiles created by the Moors are of secondary interest: it is the underlying group of symmetries which preserve aspects of the patterns that defines the geometry of the murals.

Lie also recognized that here was a kind of dictionary which translated geometry into Galois's algebraic language. Klein and Lie spent the summer of 1870 formulating their new vision of geometry until they were interrupted by the outbreak of war. France and Prussia had been rattling sabres for some months. Napoleon III saw a war as a way to bolster his flagging popularity in France. For the Prussian chancellor, Bismarck, such a clash would be a great excuse to bring the southern German states into a unified nation. A provocative letter from Bismarck ignited the tinderbox, and the French government declared war on their neighbours on 19 July.

Klein, a Prussian citizen, decided that Paris was probably not the best place to be and made a hasty exit. Lie was reluctant to leave the city since he was having such a productive time there, but when he saw how ineffective the French army was in the face of the German advances, he too departed, for neutral Italy. Lie was used to hiking across the Norwegian countryside. He had once walked the 60 km from the capital to his parents' house – only to find that they weren't in. He promptly turned around and hiked the 60 km back home. So a trek from Paris across the Swiss Alps towards Milan was something Lie looked forward to with relish.

Lie was such a hardy walker that when it rained he simply removed his clothes, put them in his backpack to keep them dry and carried on, baring all to the elements. Perhaps it was not surprising that an army unit spying a naked hiker should find him a little suspicious, and Lie was picked up by the authorities 50 km outside Paris.

When the police discovered that this strange man had a foreign accent and that his bag was full of letters written in German with pages of coded messages full of cryptic signs, it didn't take them long to conclude that he must be a German spy. Lie tried to explain that the cryptic signs were simply mathematical formulae, but the police were still suspicious. They challenged Lie to explain the theories in the papers. 'You will never, in all Eternity, be able to understand it!' he protested. But these were desperate times, when spies could be shot without too many questions being asked. He decided he had to make an effort. 'Now then, gentlemen, I want you to think of three axes perpendicular to each other, the x-axis, the y-axis and the z-axis . . .' and he launched into a description of the geometry Klein and he had been developing.

Convinced now that Lie was a lunatic as well as a spy, the police threw him into prison in Fontainebleau. For four weeks he sat in the dark, depressing confines of his cell, with only a copy of a Walter Scott novel in French and his mathematics for company. Like Galois before him, Lie found the solitude quite conducive to developing his abstract world of geometry: 'I think a mathematician is comparatively well suited to be in Prison.'

News finally got out about Lie's whereabouts. Headlines in the Norwegian press declared: NORWEGIAN MAN OF SCIENCE JAILED AS GERMAN SPY. Eventually, French mathematicians came to the rescue and visited the prison. They persuaded the guards that the scribbles were indeed abstract mathematics and nothing more suspicious. 'The sun has never seemed to have shone so clearly; the trees so green,' wrote Lie on his release.

Klein's return to Germany had gone more smoothly than Lie's attempt to flee Paris. His work was rewarded with the offer of a professorship at the university of Erlangen in Bavaria in 1872. In his inaugural lecture he took the opportunity to outline the new vision he had formulated with Lie of what geometry was really all about. The essence of geometry was captured not by the picture of the points and lines, but by the group of symmetries that permuted these points and lines. By using Galois's language it was possible to articulate much more clearly the essential components of the geometry, and to say, for example, when two geometries were the same or not. In many ways the speech was a manifesto for a new sort of mathematics, and it became known as the Erlangen Programme.

As well as formulating his overarching theories, Klein discovered another beautiful manifestation of Galois's group PSL(2, 7) as the group of symmetries of an object resembling a three-holed bagel constructed from 21 triangular faces (Figure 61). He also discovered that although the indivisible symmetries of the dodecahedron stopped a quintic equation from being solved by taking simple fifth roots, its geometry could be used to define more elaborate operations that would solve these equations. His discovery is the genesis of a completely new class of mathematical objects that are central to modern number theory. Called modular forms, they would play a crucial role in the story of the Monster symmetry group discovered a hundred years later.

Fig. 61 Klein's three-holed bagel built from 21 triangular pieces.

Lie realized that implicit in Klein's Erlangen Programme was the problem of a complete classification of what sort of groups of symmetries could possibly arise from these geometric settings. He believed at first that the problem was 'absurd or impossible', but a year later he changed his mind. He had found a way in.

Back in Norway, Lie began to formulate a whole new approach to the geometric groups that Galois and Jordan had described. He called them transformation groups, but today they go by the name of their inventor: Lie groups. It took some time for Lie's new mathematics to be recognized, partly because of the difficulty he had writing up his work: 'publication in this area went woefully slow. I could not structure it properly, and I was always afraid of making mistakes. Not the small inessential mistakes . . . No, it was the deep-rooted errors I feared.' But Lie was confident in its worth: 'my life's work will stand through all times and in the years to come, be more and more appreciated – no doubt about it'. Lie was certainly proved correct. His groups would become essential tools in theoretical physics, but they also gave rise to a whole range of new families of building blocks in the periodic table of symmetry. One of the families describes the symmetry of the hypercube Tomer and I saw in Paris last month.

A group of American mathematicians were instrumental in exploring these groups during the first few decades of the twentieth century, but it was once again a French mathematician, Claude Chevalley, who in 1954 created a framework in which all these families of building blocks, now called the simple groups of Lie type, could be described systematically. In addition to the groups Galois discovered, he accounted for another 12 different families.

In recognition of his work, Lie was offered the professorship at Leipzig that Klein had just vacated on his way to a post in Göttingen. The move, in 1886, was not a happy one. Lie had felt isolated mathematically in Norway, but the depressing dull wet weather in Germany started to get to him. He grew very homesick and wrote of how he missed the forests and mountains of Norway: 'I cannot find words to express how much I am longing to be back in Norway. My nervous system had suffered a lot here in Leipzig, where I have missed the opportunity for exercise and the spiritual influence of nature.' The hard work that Lie put into realizing his vision also began to take its toll. He suffered from insomnia, depression and stress, and in 1889 he was diagnosed with neurasthenia and admitted to a clinic in Hanover.

For seven months Lie remained in the clinic. He was an impossible patient, refusing to take the opium treatments he was prescribed. Instead he found his own way through the mist, weaning himself off the addictive drugs the doctors had been administering. By 1890 he had recovered enough to leave the clinic – 'with sleep, the pleasure of life and work has returned' – but the breakdown in his mental health had scarred his state of mind.

Lie became paranoid, convinced that people were stealing his ideas. When Klein relaunched the Erlangen Programme, Lie objected that his contribution had been edited out. Lie was further outraged when he discovered that Klein had destroyed all the letters Lie had sent him during their decade of collaboration. Lie believed that they had made an agreement to preserve the correspondence, but now the evidence of involvement had vanished. In a letter to Klein he declared that the burning of his letters was an act of vandalism. He publicly attacked his old friend in the introduction to the third volume of his book on transformation groups, published in 1893: 'I am no pupil of Klein, nor is the opposite the case, although this might be closer to the truth.' Lie cast himself as an Abel pitched against the German establishment, believing that the mathematical community were treating him as badly as they had his mathematical compatriot. As the great German mathematician David Hilbert put it, 'with the third volume, his megalomania burst into the open'.

Despite the fights over priority, Lie and Klein's mathematics established the importance of Galois's concepts to the understanding of geometry. But another perspective was being developed on Galois's work which would reveal a beautiful way to articulate why so many different looking geometries and pictures were all just the same group of symmetries dressed up in different outfits.

Applying the letter of the law

Mathematicians were beginning to understand that there were many different ways of representing the same group of symmetries. For Galois, groups of symmetries were the ways you could permute solutions to equations. But we've already seen, for example, how the permutations of the four solutions of a quartic equation can also be viewed geometrically, as the symmetries of a tetrahedron. For a gambler, shuffling a pack of cards could be described by its group of symmetries. The permutations then just correspond to shuffling the pack of cards into a new order. For Lie and Klein, the symmetries were swapping lines that ran through their geometries.

232   Finding Moonshine

But hiding behind all these different representations is a common abstraction that captures the underlying symmetry of each setting. In the same way that the generalized abstract concept of the number 8 underlies the specific physical groupings of eight stones or eight cows, Galois's group of symmetries can have many different concrete representations. Whether it is shuffling four cards, spinning a tetrahedron or permuting four lines in finite geometries, there is a common abstract entity, called the alternating group of degree 4, which captures the essential symmetry behind each example. The vision to see a common abstract mathematical concept at work in different settings took a major leap in creative thinking. This leap of abstraction was made not in the hallowed halls of the École Polytechnique or the Berlin Academy, but in rather more surprising surroundings: the Inns of Court in the heart of legal London.

By the middle of the nineteenth century, Britain had become a mathematical backwater. Newton's dispute with Leibniz over the invention of calculus had seen Britain isolated from the rest of the European mathematical community. Without contact with the major academic centres in France and Germany, Britain's mathematics began to stagnate. Newton may have been the first to come up with the mathematics behind the calculus, but Leibniz devised a far superior language in which to express the ideas. Newton was more interested in creating mathematics for himself, and less focused on communicating his ideas. His notation would vary from day to day. He had a very geometric view of the world, and this was one of his strengths, but pictures are sometimes hard to translate into a useful language. Leibniz, on the other hand, viewed calculus from an arithmetic perspective, analysing the effect of adding up smaller and smaller quantities. His work in linguistics and symbolic logic put him in a perfect position to develop such a new mathematical language. He was also a master of notation. The superior hold that this notation gave on the emerging mathematics of the calculus proved a much better springboard for the mathematical advances of the next few centuries. Indeed, the symbols and language we use today for differentiating and integrating are precisely the ones developed by Leibniz to communicate his vision.

Newton's influence on the British scientific community during the eighteenth and the early nineteenth century meant that for years mathematics in Britain was saddled with the inferior notation Newton had used. As president of the Royal Society, Newton even managed to rig an 'independent' inquiry held in 1713 into who had invented the calculus by making sure the committee consisted of sympathetic friends and then writing the final report himself. The report concluded that Leibniz had plagiarized Newton's concept of the calculus. But it was the development of a language for symmetry invented by an Englishman in the middle of the nineteenth century that started to bring England back into the mathematical fold.

The papers published in the Royal Society's journals setting out this new language were submitted not by a professional mathematician but by a successful London barrister practising in Lincoln's Inn Fields. Arthur Cayley lived in St Petersburg for the first eight years of his life. From an early age he showed a precocious talent for arithmetic and used to perform huge calculations to amuse his family, but it was the beauty of more abstract mathematics that began to intrigue him. Like many mathematicians, this talent for mental arithmetic was eventually replaced by the pleasure in pattern searching, and it is said that in later years Cayley 'was unable to count the change for a shilling'.

His love of mathematics was complemented by a great facility for languages. At Cambridge, where he studied mathematics in 1839, he also spent time learning Greek, Italian, German and French. This combination would have a significant impact on his mathematical contributions. Although he became a fellow at Newton's old college, Trinity, he did not take holy orders and so received a lower salary, insufficient to support him as a professional mathematician. So he decided instead to apply his analytic skills to practising law. He was called to the bar in 1849 and became an extremely successful barrister.

There are many character traits shared by the mathematician and the barrister. A grounding in mathematics has been the platform for some of the most successful legal careers. A successful barrister needs the ability to present a complex legal case in court, taking care to ensure that all possibilities have been covered and leaving no opening for the opposing counsel to attack. The argument is just like the process of constructing a watertight mathematical proof. Certain legal precedents become the axiomatic system within which you can twist and turn; the general theory is applied to a particular legal case. For Cayley the Inns of Court were a happy home, but his legal success was just a means to support his true passion: mathematics. During his legal career he published more mathematical papers than most mathematicians do in a lifetime of professional work.

The range of Cayley's mathematical contributions is also phenomenal. He contributed to ideas of geometry, in particular the newly emerging non-Euclidean geometries that were challenging the Euclidean axioms that parallel lines don't meet. Perhaps it was his ability as a barrister to play both prosecutor and defendant that made him quite at home moving between different geometries satisfying different axioms. But it was his contributions to symmetry and his recognition of the abstract idea hiding behind Galois's work that cemented his name in mathematical history.

Cayley's aptitude for languages allowed him to read Galois's papers, eventually published in 1846 in the French journal established by Liouville. But it also influenced his mathematical voice, which has a particularly linguistic quality to it. He was able to articulate the grammar of the language of group theory that underlies the examples Galois was using. At the time, many mathematicians found Cayley's synthetic abstract approach to mathematics difficult to penetrate, but once accepted it made Galois's groups of symmetries much easier to analyse.

A contemporary of Cayley's commented that the usual approach adopted by a mathematician in conveying his ideas was

> to take his readers by exactly the same road he had travelled himself beginning with the simple problem which first attracted attention and leading step by step to the highest results arrived at. Cayley on the contrary, usually begins by trying to establish at once the highest generalizations he has reached.

This approach of Cayley's is a very good description of the style that would come to be favoured in France during the twentieth century. His great strength was this ability to generalize. Galois had realized that you had to analyse the interactions between the permutations of solutions to equations to see whether the equation was soluble. Two permutations could be combined one after the other to give you a third permutation. The interactions between permutations of the five solutions of the quintic were sufficiently intertwined for Galois to see that the group of permutations was indivisible. Cayley's breakthrough was to say, 'Forget the equation and its solutions, just look at the interactions of the permutations.'

For example, there are six permutations of the three solutions of a cubic equation. We can write down names for these six operations: for example, let's denote by I the symmetry that leaves everything where it is, and call the other five X, Y, R, S and T. So, for example, X and Y cycle the three solutions round, while R, S and T swap two, leaving the other where it is. The interactions between these permutations can then be described by a grammar which connects these names: R followed by S is actually the same as just doing X. Cayley saw that you could tabulate these interactions in a table, capturing the essence of the symmetry.

Table 1 gives the permutations of the three solutions. The entry in the ith row and jth column is the result of performing the ith operation followed by the jth operation. Here, suddenly, is the abstraction of the underlying symmetry. If instead I take the symmetries of a triangle and give them the same names (taking care to match them up in the right way), I find that the combinations of the symmetries of the triangle obey the same rules as those described in the table (Figure 62). For example, X matches the symmetry that rotates the triangle anti-clockwise by a third of a turn, and Y with the rotation in the opposite direction. R, S and T match the reflections in the lines of symmetry running through A, B and C, respectively. The symmetries of the triangle interact in exactly the same way as the shuffles of the three solutions to the cubic. For example, doing R followed by S is again the same as just doing X.

Cayley's language provides a way to express the fact that the group

	I	X	Y	R	S	T
I	I	X	Y	R	S	T
X	X	Y	I	T	R	S
Y	Y	I	X	S	T	R
R	R	S	T	I	X	Y
S	S	T	R	Y	I	X
T	T	R	S	X	Y	I

Table 1

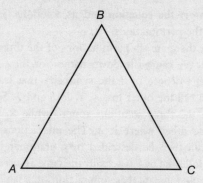

Fig. 62 The triangle has six symmetries.

of symmetries of a triangle is the same as the group of symmetries of
permuting three solutions of a cubic equation. Even more powerfully,
it finally pins down why the group of six symmetries of a triangle is
different from the group of six rotational symmetries of the six-pointed
starfish we encountered in Chapter 1 (Figure 63). Although the
rotational symmetries of the starfish can be described by six letters,
the interactions give a completely different table. In Table 2, the letters
I, X, Y, R, S and T stand for the rotations of the starfish by zero, one,
two, three, four and five sixths of a turn, respectively.

	I	X	Y	R	S	T
I	I	X	Y	R	S	T
X	X	Y	R	S	T	I
Y	Y	R	S	T	I	X
R	R	S	T	I	X	Y
S	S	T	I	X	Y	R
T	T	I	X	Y	R	S

Table 2

Thanks to Cayley, you could now say that the group of symmetries
of a tetrahedron or the shuffles of four playing cards or the permutations of the four solutions of a quartic equation were all manifestations of the same group of symmetries. Similarly, the even shuffles

Fig. 63 The six-pointed starfish has six rotational symmetries.

of five cards is the same group of symmetries as the rotations of a dodecahedron, which is the same as the permutations of six lines in Galois's geometry based on the prime 5.

In fact a group of symmetries – or simply just a 'group' – was really just a set of names, A, B, C, . . . , and a table indicating how they interacted. The table defines a kind of multiplication between the symmetries. Cayley realized that there were certain rules that a table of interactions would have to satisfy before it could define the multiplication law of a group of symmetries. There were limits on how the A, B, C, . . . could be arranged in the table: for example, rather like a sudoku, each letter had to appear once and once only in each row and in each column. These rules, or axioms, that Cayley said were necessary for a table to define a group were remarkably simple. Yet they captured in a new language the essence of symmetry. For example, the Tables 1 and 2 are the only tables that satisfy these axioms when there are six symmetries. So there are just two groups with six symmetries.

Here was a way to explain, for example, why two completely different sets of tiles on the walls of the Alhambra were actually two different expressions of the same symmetry group. If you write down names for each of the symmetries on each wall and tabulate their interactions, you get the same table for both walls. Just as Cayley's axioms tell us that there can be only two different tables with six symmetries, the language of group theory gives us the means to prove that 17 – and no more – different symmetry groups are possible on a two-dimensional wall. Cayley's vision was actually so ahead of its time that it would take until the end of the nineteenth century before mathematicians were sufficiently fluent in this new language to prove this result.

Almost simultaneously, yet independently, three mathematicians – the German Arthur Schoenflies, the Russian Eugraf Fedorov and the Englishman William Barlow – confirmed that the Moors of southern Spain had not missed an 18th way to tile their great palace. They went on to show that if one moves from two to three dimensions, then there are 230 ways to fill space with bricks rather than tiles, where the bricks form repeating three-dimensional units. This is of extreme importance in crystallography, because it means that any crystal must have a structure that corresponds to one of these 230 different symmetries.

Cayley's papers appeared in print in 1854, while he was still at the peak of his legal career. In 1863 he was offered the Sadleirian Professorship in Cambridge. He had by this time earned enough money that the cut in salary he suffered by accepting the position was not too painful; the opportunity to devote all his time to his first passion, mathematics, was payment enough. By the end of his career he had published over 900 papers. He was held in great regard by his followers in Cambridge 'with reverence almost akin to worship', as one obituary put it. A contemporary, George Salmon, summed up the contribution Cayley made to mathematics:

> The knowledge which mathematicians now possess of the structure of algebraic forms is as different from what it was before Cayley's time as the knowledge of the human body possessed by one who has dissected it and knows its internal structure is different from that of one who had only seen it from the outside.

Cayley had woken the dormant mathematical forces in Britain. One of his students, William Burnside, took up his mantle. Galois had discovered that in addition to the prime-sided shapes there were new indivisible groups, such as the alternating groups and the Lie groups. Despite the introduction of these new indivisible groups, Burnside's work revealed that, remarkably often, a group of symmetries is built from just the prime-sided building blocks.

Burnside had gone up to St John's College, Cambridge, to read mathematics. His other passion was rowing, and he was really keen to row in the college's first eight. But St John's had an excellent collection of oarsmen, and Burnside knew that his lack of weight would make it hard for him to compete for a place. So he decided to move colleges, and enrolled at Pembroke. Here he was more successful, and rowing at number seven he captained Pembroke's first eight.

At Pembroke, Burnside came into contact with a group of applied mathematicians who were studying problems in hydrodynamics, and this influenced his early mathematical tastes when he started doing his own research. It was not until he reached his early forties, when he accepted a professorship at the Royal Naval College in Greenwich, that he started his exploration of symmetry groups. One of his first achievements in 1893 was to show that the alternating group of degree 5 with 60 symmetries was the only possible building block, or simple group, with 60 symmetries. If you tried to write down a multiplication table for another group with 60 symmetries, distinct from the rotations of a dodecahedron, you would always find that it could be broken down into smaller symmetries. There are in fact precisely 12 other groups of symmetries with 60 symmetries in addition to the alternating group.

Burnside found that he was quite skilled at proving what was possible just from knowing how many symmetries there were in a group. His primary focus was groups of finite order, that is groups with only a finite number of symmetries. His greatest breakthrough, in 1904, was to show that whenever the number of symmetries was divisible by at most two primes, then the group had to be built from the simple prime-sided shapes. For example, a group with 1,000 symmetries cannot be indivisible, because $1,000 = 2^3 \times 5^3$ and it is therefore built from the rotations of three pentagons and three flips. Even a group with a googol number of symmetries (a googol is a number written as 1 followed by 100 zeros) can be broken down into the rotations of 100 pentagons and 100 flips.

Burnside could see there was still some work to do to catch up with the advances being made in Europe and the United States. In his presidential address to the London Mathematical Society, he declared:

It is undoubtedly the fact that the theory of groups of finite order has failed, so far, to arouse the interest of any but a very small number of English mathematicians; and this want of interest in England, as

compared with the amount of attention devoted to the subject both on
the Continent and in America, appears to me very remarkable.

To spur his contemporaries on, Burnside wrote a book on the theory
of groups which inspired many mathematicians, and not just in England, to dive into this new mathematics. Many agreed that he had
succeeded in the task he set himself in the introduction: 'It will afford
me much satisfaction if, by means of this book, I shall succeed in
arousing interest among English mathematicians in a branch of pure
mathematics which becomes the more fascinating the more it is
studied.'

Burnside's theorem on the divisibility of groups whose order is $p^a q^b$
proved to be just what was needed to identify the simple groups
with a small number of symmetries. For example, apart from the
prime-sided shapes, the only simple groups with at most 200 symmetries are the rotations of a dodecahedron (with 60 symmetries) and
Galois's permutations of the lines called PSL(2,7) (with 168 symmetries).

In the second edition of his book, published in 1911, Burnside used
this theorem to identify all the simple groups with fewer than 1,092
symmetries. But he had a hunch that his theorem encompassed much
more than just groups whose number of elements were divisible by
two primes. He believed that if the number of symmetries was odd,
then the symmetries could always be broken down into the simple
prime-sided shapes. If this were true, it would be a major advance
along the road to a complete classification of the building blocks of
symmetry, for it would immediately wipe out half the groups that one
would need to consider. And knowing that a building block has to have
a number of symmetries divisible by 2 means that one of the symmetries
must look something like a reflection, and this might provide a real foot-
hold into a complete analysis of these building blocks.

In the decades that followed, mathematicians became more optimistic that if Burnside's theorem could be proved, it would help to wrap
up the project of classifying the simple groups. The building blocks of
symmetry would then consist of the prime-sided shapes, Galois's even
shuffles (the alternating groups) and all these wonderful geometric
families, 13 in total, that Lie's work revealed but whose genesis still
can be found in those memoirs of Galois.

There was just one minor blot on the aesthetics of such a beautiful classification. In 1860 the French mathematician Émile Mathieu had discovered five rather strange shuffles which produced indivisible symmetries. These five simple groups didn't seem to fit into any of the known families, nor did they seem to create an infinite family of their own. They were just five rather strange groups of symmetries, indivisible but not fitting into any obvious pattern of groups. These groups would turn out to be just the thin end of the wedge. But the discovery of the rest of the wedge would take nearly another century and a detour through the application of symmetry to the telecommunications industry.

28 March, Stoke Newington

It's not ideal, but I think it will have to do. Burnside was showing how to break groups down into prime-sided shapes; I've spent the month writing a paper on how to put these bits back together. The paper is a first step towards trying to see how the number of groups you can build from prime-sided shapes varies as you change the prime number. For example, there are only ever two symmetrical objects you can build with p^2 symmetries, where p is any prime number. The number of symmetrical objects doesn't depend on the prime p.

When you consider the number of groups with p^3 symmetries, then there are always five symmetrical objects you can build. For groups with p^4 symmetries, if the prime is odd you get 15, and if it is even, i.e. $p = 2$, then you get only 14. But when you move to p^5 symmetries, things start to get interesting. The number of symmetrical objects with p^5 symmetries now depends on the prime: as the prime gets bigger, so does the number of symmetrical objects you can build. The number of objects is essentially got by doubling the prime. Add another prime, and the number of symmetrical objects with p^6 symmetries is given by a quadratic polynomial in the prime p.

The big question that I would love to answer, the PORC Conjecture, is whether, as you increase the number of shapes you are using, there is always a simple equation that gives you the number of symmetrical objects. It is not clear that this will always be the case. Perhaps the number of groups with p^{10} symmetries will be controlled by some different sort of mathematical function. My elliptic curve example discovered in Bonn suggests that it's possible that at some point the answer might depend on having to know the number of pairs of numbers (x, y) that make $y^2 - x^3 + x$ divisible by p. This number isn't given by a simple formula.

I must admit that at this point my mind is completely open about which way this problem might go. It's quite exhilarating not to know. The paper I've been writing is working towards showing that if you add certain extra conditions about how the prime-sided shapes fit together, then the number of symmetrical shapes you can build in this particular way is given by a simple equation. The ultimate aim will be to see what happens if you don't force the shapes to have to be put together in this special way.

Big theorems are like jigsaw puzzles. You can't hope to do the puzzle in one big go – it's a gradual cumulative effort, sometimes involving several people. Nevertheless, who wouldn't enjoy being the person to put in the last piece? That's why I'm slightly reluctant to send the paper off in its current state. But in the academic world there is a lot of pressure to keep pumping out publications – 'publish or perish' – and I will have to submit the paper to a journal before knowing whether I can finish the puzzle.

An email has just arrived which offers me a welcome break from the hard work. Dorothy Ker, a friend of mine who is a composer, suggests that we meet up next week. Over the last few years we have been meeting to discuss connections between mathematics and music. As mathematicians were beginning to grapple with the abstract ideas of symmetry, musicians of the early twentieth century were looking to formal structures and mathematics to replace the tonal heritage they had cast aside. Even early classical music plays lots of mathematical games to generate interesting variations on themes. In fact, the amount of symmetry that I've discovered hidden away in musical compositions makes me wonder whether this is why I find music such an ideal accompaniment to creating mathematics. I put on some Bach to see if it will help to inspire me.

April: Sounding Symmetry

Cascades in music, gentlest of all time's shapes.

JORGE LUIS BORGES, 'Matthew XXV:30'

5 April, London Bridge

Once nineteenth-century mathematicians had released symmetry from physical objects, people started to find it in the most unexpected places. For me, one of the most interesting abstract expressions of symmetry is in music. There have always been close bonds between the two disciplines. Leibniz once declared that 'Music is nothing but unconscious arithmetic.' But the connection extends beyond simple counting and rhythm.

The structures that musicians enjoy threading through their music have a distinctly mathematical flavour. Certainly many musicians are conscious of the connection. The French Baroque composer Jean-Philippe Rameau was one such: he wrote in 1722 that, 'Not withstanding all the experience I may have acquired in music from being associated with it for so long, I must confess that only with the aid of mathematics did my ideas become clear.' It would turn out that composers had for centuries been toying with the concept of symmetry, but a complete understanding of what they were doing became possible only with the new mathematical language developed by Galois.

I have fantasies about what I might have done with my life had I not become a mathematician: running away to train at the Le Coq theatre school in Paris, setting up my own restaurant. Becoming a composer is also up there with my theatrical and gastronomic aspirations, which is why I get some vicarious pleasure from my occasional meetings with Dorothy. She uses a lot of natural and mathematical structures as a framework or launch pad for composition. I'm intrigued as to how this mix of mathematics and music works. For her part, Dorothy is keen to explore other less obvious structures and to get a sense of my mathematical way of looking at the world.

We've arranged to meet up at London Bridge this afternoon. It's a bright day and the Thames is at low tide, so we head for a small patch of sand on the riverbank, where the mudlarks used to go hunting for treasure. It's interesting that our conversations are never very fluid affairs. Both of us are used to expressing ourselves in non-verbal media. When I'm nervous I have the annoying habit of filling silence with any ideas that come into my head. When I'm with Dorothy I try to shut up and not burble.

We find a lot of similarities in our working practices. I've spent the last few weeks painstakingly trying to write the proof of the theorem I proved with Fritz some months ago. Dorothy's spent the morning doing something very similar. For a musician, the hard labour is writing out the detailed score of a piece. The musical ideas might have come during a stroll in the hills or a train journey, but the majority of Dorothy's work is translating those ideas into notes on the page for all the various different musical instruments. Intriguingly, we both find being in motion an important ingredient in stimulating those flashes of inspiration.

We start by looking at a piece Dorothy has written for solo cello that uses the Fibonacci sequence, 1, 1, 2, 3, 5, 8, 13, 21, . . . , as a skeleton. This sequence is a favourite with artists because of its strong connection with growth. The way it starts from something tiny and evolves with increasing complexity resonates with the development of a musical work. Dorothy has brought along a recording of the cello piece and a copy of its score. We listen to it on her portable CD player. Without insider knowledge, I don't think I'd have spotted the Fibonacci sequence. But knowing it's there does increase my enjoyment of the piece.

I can see that Dorothy's notepad is filled with lots of numbers arranged in what looks a matrix. This is the mathematical object that Galois used to swap his lines around in his finite geometries. In the mathematician's lexicon, a matrix represents a symmetry. Dorothy also calls it a matrix, but when I ask her whether it moves anything round she looks a little confused. It turns out that the musical matrix is really just a table with 48 rows for keeping track of musical themes.

The grid looks like a huge sudoku puzzle. In each row, the numbers from 1 to 12 appear once and once only. Each number represents one of the 12 notes in the chromatic scale. The first row indicates a choice of theme. There are a total of

$$12 \times 11 \times 10 \times 9 \times 8 \times 7 \times 6 \times 5 \times 4 \times 3 \times 2 \times 1 = 479,001,600$$

different themes to choose from. Each subsequent row is a variation on that first theme. The first row is the seed from which Dorothy then grows a garden of sound.

Particularly intriguing is the fact that the rules for this growth are various tricks of symmetry that you can play on the original theme. The first is simply to reverse the order of the notes – what musicians refer to as the retrograde. We can view this variation geometrically by looking at the pattern of notes on the musical score: the notes on the stave are reflected in a vertical line running through the middle of the music. The second symmetry is essentially a reflection in the horizontal line. If the original theme climbs up three notes, for example, the variation descends three notes. The third symmetry combines the first two. Dorothy takes these first four rows – the theme and its three variations – and translates the pitch systematically through the 12 different shifts of the chromatic scale.

There is a definite symmetrical object I can identify that would capture the 48 different possible musical threads that Dorothy has generated. The reversing and inverting are actually the same as the symmetries of a rectangle, while the 12 shifts of the notes are the same as the rotational symmetries of a 12-sided coin. The label I'd put on this group of symmetries is $C_2 \times D_{12}$, or to spell it out, 'the direct product of the cyclic group of order 2 with the 12th dihedral group'. Dorothy tells me that musicians are particularly drawn to those elements of the original theme that remain unaffected by the twists and turns, rather like those symmetries in the walls of the Alhambra that also match up colours. The symmetrical object behind the transformations has provided a palette of sound from which Dorothy then starts her composition, the artist taking over from the mathematician.

The system Dorothy is using was introduced by Arnold Schoenberg, the father of the modern obsession with symmetry in music. The interest in mathematics, and in symmetry in particular, as a framework for musical composition became quite strong at the beginning of the twentieth century, when composers rejected their tonal heritage and looked for other ways to give structure to their compositions. Schoenberg more than anyone pushed to the limit this application of mathematical transformations to musical themes. Many people listening to the music of Schoenberg for the first time hear only the sound of chaos, but there is actually a lot of structure at its heart.

Schoenberg's methods influenced many of his students. In Alban Berg's *Lyric Suite*, the 46 bars that open the third movement are, after another 23 bars, mirrored in the concluding 46 bars. The numbers 23 and $46 = 2 \times 23$ are no accidents, either. Berg regarded 23 as his signature number. Rather like a footballer being identified by his shirt number; if you hear 23 in music, you know it's Alban Berg. Anton Webern, Olivier Messaien and Pierre Boulez were other composers who used Schoenberg's symmetrical methods.

Dorothy once told me of the risks that a composer runs in relying too heavily on such generative structures:

> Perfect symmetry realized without human intervention can be bland and lacking in tension. A process that is too obvious trails far behind the listener's ability to predict its outcomes. Such music – to borrow the words of Harrison Birtwistle – 'finishes before it stops'. The music we value most seems to be that which succeeds in achieving a 'perfect' integration between image and form.

I make very similar aesthetic judgements in singling out the mathematics that I appreciate. The mathematics that is hard to predict is what interests me the most. The best mathematics, in my view, is what, despite the rigid constraints of the formal logic, still produces moments full of surprises.

Symmetry in 32 movements

The modern use of symmetry in music has old roots. One of the first to exploit these ideas was the master of using mathematics in music, J. S. Bach. His *Goldberg Variations*, published in 1741, is perhaps one of the best examples of the sound of symmetry. I've been plodding my way through its 32 movements in my breaks from mathematics, exploring the mathematical tricks Bach has exploited to create his variations.

Each movement consists of 32 bars repeated twice. The piece begins and ends with a simple yet elegant aria which establishes the motif that Bach proceeds to twist and turn over the 30 variations. By repeating the aria at the end, he evokes the shape of one of the most symmetrical of objects: the circle. The repetition of the aria connects the two ends of this musical braid. With the 32 movements in a circle, the 16th variation sits diametrically opposite the first rendition of the aria. Interestingly, Bach refers to the 16th variation as the overture, a term usually reserved for the opening of a piece of music. It makes one begin to question exactly where the circle begins and ends.

The 30 variations are arranged as ten groups of three, and the third movement in each group is a canon. It is in the bass line, which consists of four phrases of eight notes, that Bach plays most of his symmetrical games – you can see and hear the music being stretched, squeezed, reflected and sent spiralling through the 30 variations.

It is within the cycle of canons that Bach really exploits the potential of symmetry as a source of variations. A canon by definition is an example of translational symmetry. This is the kind of symmetry produced by sliding a copy of a shape with respect to the original, like a decorative motif being repeated in a 'frieze' pattern around the top of a pot. But the musical version is a temporal translation rather than a spatial one – a shift in time. One voice starts singing a phrase, and then after a few notes a second voice comes in and sings the same phrase over the top of the first voice. Examples of these songs, also known as rounds, are 'Frère Jacques' and 'London's Burning'. Dorothy once pointed out a modern example: anyone who has listened to a radio broadcast simultaneously online and on an analogue radio will have effectively created a canon. The time taken by the computer to process the digital signal results in the music being played a few seconds behind the broadcast via the analogue radio. When you look at the score of the *Goldberg Variations*, you see the temporal translational symmetry rendered as a spatial one: the same sequence of notes is repeated several beats on from its first occurrence, just like a pattern round the top of a pot.

Bach, though, is not satisfied with simple translation in the temporal dimension, and starts to play the same trick in the pitch dimension. In the second canon (variation number six) the second voice starts one note higher than the lead voice. The effect is now like a diagonal motif spiralling up the side of a pot. Each subsequent canon moves the second voice up one more note. There is a larger difference in pitch between the two parts each time we hear a new variation. But then something quite striking happens.

By the time we reach the eighth canon we suddenly feel the two voices have come together again. There is an octave difference between the voices. The beautiful thing about the octave is that our brain senses an identity between these two notes, something Pythagoras discovered two millennia before Bach's use of the octave to complete the circle. It feels like one of Escher's paradoxical pictures where the monks climb a quadrangle of steps only to find themselves at their starting point again.

Interestingly, Bach starts to climb again with a ninth canon which stretches the ear further as the second voice now starts an octave and a note higher. Like a corkscrew, the canons seem to want to spiral up endlessly, every eighth canon aligning with the first canon as we hear another octave. The idea of themes spiralling up in pitch throughout a piece has been used by many composers. In his opera *The Turn of the Screw*, Benjamin Britten uses the spiralling shift in pitch throughout the piece to express the tightening of the screw as the ghost gains more of a hold on the young Miles, until the piece climaxes with the boy's death.

The *Goldberg Variations* is like a musical version of a torus. The torus is a mathematical object generated by a circle sweeping out another circle (Figure 64). So you've got circles running horizontally but also circles running vertically. The shape is like a circle of circles. In the *Goldberg Variations* we hear circles in time and circles in pitch.

What should be the tenth canon actually breaks the structure. It is

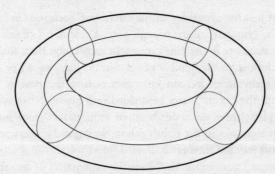

Fig. 64 The torus – a circle of circles.

called a quodlibet – a kind of musical joke – and is a contrapuntal piece based on two common folk tunes of the day. Perfect symmetry for many artists is something unsettling. It is by its very nature pre-scriptive: once you understand part of the symmetrical structure, you know what is going to come next. Bach seems to have felt a need to break the symmetry and weave a different variation as his climax. But breaking the symmetry also highlights just how much pattern has gone before.

Breaking symmetry in art has a long heritage. The carpet weavers of Persia and Arabia would deliberately weave a fault into a small portion of their otherwise beautifully symmetric carpets to destroy the perfect balance. They believed that their souls were woven into the carpet, and that by including some imperfection they were leaving a route for their souls to escape. Perfect symmetry for some Muslims also meant trying to imitate God – an act of great blasphemy. By including a fault in the pattern they wouldn't challenge God's position as the master weaver.

Even today, in West Africa, the symmetry of weaving is considered the key to binding supernatural power into the piece. Repeating dia-monds that get smaller and smaller towards the centre are used as motifs in the designs for wedding blankets made by the Fulani people of Mali. They believe that, with each repetition of the pattern, more and more spiritual energy is being sealed into the blanket. Indeed, the energy bound up in the blanket is regarded as potentially so dangerous that the engaged couple are supposed to make sure that the weaver stays awake all night lest he fall asleep and release the force bound in the threads.

It's not just the symmetry of the circle that Bach uses in his variations. The symmetry of the triangle plays a role in his choice for the rhythmic structure in the nine canons. In each canon Bach makes one of three choices for the number of beats in the bar: two, three or four. The other rhythmic decision he makes is how the beat is divided throughout the canon. It can be divided into quavers (two notes per beat), triplets (three notes per beat) or semiquavers (four notes per beat). For example, in the eighth canon, variation 24, there are three beats in the bar and the beat is divided into triplets.

As soon as I understand this, I see the symmetries of two spinning triangles at work in determining the rhythmic structure throughout the piece. The structure in each canon can be interpreted as a symmetry of the combination lock with two spinning triangles (Figure 65). In the lock, each symmetry is a way of spinning the two triangles so that a different combination of numbers appears on the side. But instead of numbers on the sides of the triangles, I can put rhythmic ideas. The first triangle controls the number of beats in the bar, so the outfacing edge of the triangle will indicate two beats, three beats or four beats. The second triangle keeps track of the division of the beat, and so will indicate quavers, triplets or semiquavers.

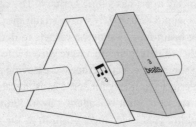

Fig. 65 The rhythmic choices for each canon correspond to the symmetries of spinning two triangles.

The wheels are then spun through the different combinations, and whatever rhythmic idea appears on the outfacing edge of each triangle is used in the next canon. The musical triangles spin around as if Bach is cracking a combination lock with two wheels. In total there are nine symmetries of the combination lock. Bach's mastery is to systematically go through each of the nine symmetries or combinations and assign one to each of the nine canons. As he spins the triangles throughout the composition, no possible combination is missed.

Not content with playing symmetrical games with rhythm, pitch and time, Bach has even more symmetrical tricks up his sleeve. In the fourth canon the second voice is not a simple replica of the first but is inverted – so when one voice climbs up, the second voice descends by the same pitch. On the score, the notes are simply flipped or reflected in the horizontal line running through the music. The same reflection is used in the fifth canon.

It is striking that the standard Western musical notation turns music into something geometric where you can actually see all these abstract symmetries. The symmetry in the sound gets translated into a visual symmetry. Symmetrical games in the vertical direction translate into transformations in the pitch of the music. Using symmetry in the horizontal direction affects the music's temporal structure. There are other variables which the composer can play with, such as the loudness of the music. The symmetry of a crescendo followed by a diminuendo in a piece translates into seeing the symbols $<$ and $>$. So a piece of music can capture symmetry happening in many dimensions – time, pitch, rhythm, volume.

One of the obvious geometric symmetries that was not exploited until the twentieth century is the idea of rotating themes. Paul Hindemith tries this trick in his piano piece *Ludus Tonalis*, or 'Game of Tones'. Take the notes written at the beginning of the piece. Turn the score upside down so that the notes on the page are rotated by half a turn, and you will get the notes that conclude the last movement. Admittedly there is an hour of music in between, so Hindemith is hardly expecting you to hear the symmetry. Yet it's no coincidence that it's there. It acts like a literary device called chiasmus, which uses symmetry to denote the beginning and end of a text like a pair of bookends.

A few years ago the BBC ran a competition where you had to listen to three recordings and decide which was a pastiche written by a modern composer, which was genuine Bach and which was computer generated. The fact that such a competition is possible says much about the high degree of structure in Bach's compositions. The *Goldberg Variations* is a musical journey through the world of symmetry.

Listening to it is like walking through a hall of mirrors, as each new variation twists and stretches the original theme. Bach's student Lorenz Mizler referred to music as the process of 'sounding mathematics'. The *Goldberg Variations* is a good example of how symmetry is not just a physical property but pervades many abstract structures.

Given all the symmetry in the *Goldberg Variations*, one might expect Bach to be my favourite composer. But perhaps the pervasive symmetry in the *Goldberg Variations* explains why the piece has never excited me in the way I know it is meant to. It is probably sacrilege to admit to it, but even Glenn Gould playing the variations doesn't get my blood racing like listening to a piece by Richard Strauss can. Maybe the predictability that symmetry provides is killing the sense of surprise that I'm after. When I confess this to Dorothy, she reminds me that the *Goldberg Variations* was written to help Count Hermann Karl von Keyserling fall asleep. Dan, the mathematician in Oxford who supervised my mathematical research, says that I'm just still too young to appreciate the *Goldberg Variations*. Perhaps I'm not yet old enough to see the ultimate symmetry embedded in the piece – the reflection of the human condition. For my 40th birthday he gave me a recording of Bach's *Well-Tempered Clavier*, saying, 'Maybe you are now old enough at 40 to appreciate these.' I'm still working on it.

Pattern searching

Bach often used symmetry as a short cut to save writing out a score in full. For example, in the fourth and fifth canons of the *Goldberg Variations*, the second voice is a horizontal reflection of the opening phrase. To indicate that the second voice needs to be played upside down, Bach included a second clef at the beginning of the piece, but he inverted it (Figure 66).

This practice of notating the theme with additional instructions to create variations had been used in much of the polyphonic music written by Bach's predecessors, such as Josquin des Prez. Each performer would be given the same single line of music, but with their own set of instructions on how high to translate the pitch, how long to wait before entering and how the speed was being dilated or contrac-

Fig. 66 The inverted treble clef indicates that the theme is inverted
when it is heard the second time.

ted. To generate their part of the composition, each performer would
carry out these instructions on the basic musical line like someone
performing symmetrical moves on a physical object.

One of Bach's most overt ways of incorporating symmetry was to
use a form known as a crab or retrograde canon, which he did in
another collection of pieces known as the *Musical Offerings*. In a crab
canon, one voice plays the part in the usual way, from start to finish,
while at the same time the second voice starts at the last note of the
piece and plays it backwards. The art of the composer is of course in
writing a single line that can be played in this crab-like fashion. The
crab is like a musical palindrome. Some musicians have composed
truly palindromic pieces, where halfway through the piece you play
the music you have just played but in reverse. The minuet from
Haydn's Piano Sonata No. 41 is a perfect palindrome. In the interludes
of his opera *Lulu*, Berg used palindromic musical structures, as did
Béla Bartók in his Fifth String Quartet. Wolfgang Amadeus Mozart
also enjoyed exploring palindromic music.

The highly structural nature of early music is behind a well-known
story about Mozart's precocious musical abilities. During Holy Week,
the matins service in the Sistine Chapel would conclude with a performance of Gregorio Allegri's setting of the *Miserere*. Written originally for Pope Urban VIII, the piece would be sung as 27 candles were
gradually extinguished to leave one candle burning. As the castrati
singers soared off into the heights, the Pope himself would fall to his
knees in front of the altar, dramatically ending the service. The Pope
loved the piece so much that he decided to keep it for the sole use of
the Vatican. A Papal decree forbade the performance of the *Miserere*

other than during Holy Week in the Sistine Chapel. No manuscripts were allowed to leave the Vatican, and excommunication awaited anyone who tried to make a transcript.

In December 1769 the 14-year-old Mozart set off on a tour of Europe with his father. One of the highlights was attending the famous Holy Week matins service at the Vatican to hear the *Miserere*, the one chance in the year to listen to the beautiful piece. The boy was so taken by the performance that when he got back to his lodgings that evening he sat down and wrote out the complete transcript of the 12-minute piece from memory. Although risking being excommunicated, he furtively returned for the Good Friday performance to check how accurate his manuscript was. Needless to say, it required only a few minor corrections.

The act of recreating the piece wasn't so much a feat of memory as a reflection of Mozart's extraordinary ability to understand the inner logic of the composition. Because of the patterns and symmetry running through the work, it provided Mozart, perhaps subconsciously, with an algorithm to rebuild the nine-part choral piece. It is an almost impossible task to remember a random string of numbers such as 99375105820974944592, but not so for a string such as 12345543211234554321. This second string has a symmetry, which enables your brain to store the 20 numbers by using a program, and this is less demanding of brainpower than having to commit to memory a random string of 20 numbers. The same principle is at the heart of Mozart's impressive musical memory. He was highly sensitive not to numerical sequences, but to musical patterns.

The young Mozart's exceptional musical gift gave him the insight to deconstruct Allegri's piece according to its inner symmetry. In a similar way, someone with great mathematical insight might spot that the 20 random numbers in the previous paragraph are the 44th to 63rd numbers in the decimal expansion of π. Like John Conway's ability to recall π to thousands of decimal places, Mozart's reconstruction of the *Miserere* is the sign not of an amazing memory but rather of a mind sensitive to the symmetry and patterns threaded through the piece. Memory, both in the human brain and in computers, is connected very often with an ability to spot structure or connections which allow the hardware to store information in compressed form.

What Mozart did with Allegri's *Miserere* is connected with what I'm ultimately trying to do with my sequence of numbers describing the number of symmetrical objects you can build from the symmetries of the triangle. By understanding the logic and pattern at the heart of the *Miserere*, Mozart could reconstruct the whole piece just from fragments in his memory. In my mathematical composition, I've got the first phrase as 1, 2, 5, 15, 67, 504, 9,310, . . . , but I don't know how it will continue. I've not yet found the secret behind the melody.

While Mozart was performing his music for the courts of Europe, a German physicist was impressing them with a different way of seeing symmetry in music. Born in the same year as Mozart, Ernst Chladni discovered how one can *see* the sound of a drum. By placing sand on the surface of the drum and vibrating the skin, he was able to produce an extraordinary range of patterns in the sand, full of symmetry.

These shapes are similar to the waveforms of the various harmonics of a violin string. When a violin string vibrates it is actually combining all the different sine waves that fit into the length of the string. By placing a finger lightly at different points on the violin string, it is possible to pick out the harmonics that make up the sound of the violin. For example, placing a finger halfway along the string picks out the first harmonic, a note an octave higher than the fundamental note of the violin string. The second harmonic, which sounds a perfect fifth higher, is got by placing the finger one-third of the way along the string. What Chladni discovered is that a drum also has versions of these harmonics, but instead of waves on a one-dimensional string one finds amazing two-dimensional shapes appearing across the surface of the drum (Figure 67). Each different shape is got by a similar process to placing your finger at different points on the violin string. It is the combination of all these different patterns and their corresponding frequencies that makes up the characteristic sound of each drum.

His demonstrations were so successful that Chladni toured the courts of Europe exhibiting the symmetry hidden in the sounds of different musical instruments. Napoleon was particularly keen on Chladni's travelling show and rewarded him with a handsome gift of 6,000 francs. We now understand that these symmetries are at the root of the different tonal qualities of a cheap violin and a Stradivarius. The German violin makers were specialists in crafting instruments to

Fig. 67 Ernst Chladni discovered that the vibrations of a drum are full of symmetrical patterns.

produce as much symmetry as possible in the sound waves that vibrate inside the box.

The musicians of the baroque and classical periods were actually coming to grips with the abstract nature of symmetry well before mathematicians really understood the concept. For these musicians, symmetry was already more than geometric reflections and rotations. One group of musicians even unwittingly proved some of the first mathematical theorems about symmetry.

Ringing the changes

A few centuries before Galois's and Cauchy's mathematics of permutations, the bell-ringers of seventeenth-century England had, without
knowing it, already started to prove quite complex mathematical
theorems about the theory of permutations. They had discovered that
by repeating a sequence of simple permutations they could generate
every permutation possible. But lacking the sophisticated nineteenth-
century language of groups of permutations, they were not aware of
the interesting mathematical theorems that they were putting into
practice every Sunday.

Living in Oxford as a student, I got quite used to the beautiful sound
of pealing bells. It is one of the characteristic sounds of English life
and has been familiar to students walking the streets in Oxford for
nearly four centuries. There are over 5,200 sets of bells hung in church
towers across England, most containing five or six different bells but
some with as many as a dozen.

As I made my way to lectures on permutations, I passed the ten
bells ringing in the church of Mary Magdalene, but had no idea that
the bell-ringers were putting into practice the theory I was about to
learn. Inside the bell tower, ten bell-ringers clung to ten ropes dangling
from the belfry. It is a feat of coordination to get the ten bells to ring
in sequence one after the other, from the highest-pitched bell to the
lowest. But things can get pretty boring once you've got that sussed.
So just as centuries of bell-ringers before them had done, they began
to ring the changes.

Seventeenth-century bell-ringers started to play around, varying the
order of the bells. For example, suppose there were just four bells,
called A, B, C and D. The bell-ringers start ringing them in the sequence
ABCD. When the conductor shouts 'Change!', the order in which the
first and second bells are rung is swapped, and the third and fourth
are swapped too. So after one change, the order in which the bells are
rung is now BADC.

If at the next change the same pairs swap, B with A and D with C,
the sequence BADC returns to ABCD, the original order. So bell-ringers in
the seventeenth century started experimenting with trying a different
way of swapping the bells when the conductor shouted 'Change!' for

the second time. Changes in bell-ringing are possible only if two people
ringing bells one after the other swap in the following sequence. So
the second change might swap the second and third bells being rung.
At the call of the second change the ringers of bells A and D, the
second and third in the sequence, swap round, and BADC becomes
BDAC.

When the conductor calls for the next change, the bell-ringers make
the first change again: the first and second change their order, as do
the third and fourth. So now BDAC becomes DBCA. We already have
four different sequences of bells. If we continue we get eight different
sequences before we get back to the beginning. With four bells it is
possible to get 24 different sequences: there are four choices for the
first bell, for each of which there are three choices of what to play
second, then a choice of two bells for the third to be sounded, and the
bell that's left goes last. So that's $4 \times 3 \times 2 \times 1 = 24$ different sequences.

We've managed to generate eight of these 24 possible sequences by
using the two changes described above, one swapping the first and
second bells and the third and fourth bells, the other swapping the
second and third bells. The sequence of bells created by these two
changes is the start of a sequence called Plain Bob Minimus. Bell-
ringers took up the challenge of how to get all 24 sequences. This is
possible if after eight sequences the conductor adds a third change
where just the third and fourth ringers swap.

To see where the symmetry is hiding in this bell-ringing, imagine
the bell-ringers on the corners of a square. Place the square in an
outline and number the corners 1, 2, 3 and 4, as in Figure 68. Each time
the conductor calls a change, the bell-ringers respond by exploiting a
symmetry of the square. After the letters flip around, the bell at corner
1 rings first, the bell at corner 2 next, and so on. The first change
corresponds to a reflection in the horizontal axis and the second
change is a reflection along the diagonal through corners 1 and 4. The
combination of these reflections is enough to produce all the eight
different symmetries of the square. Galois's language allows us to say
that although four bell-ringers performing the opening eight rounds
of Plain Bob Minimus and the geometry of a square appear very
different, the symmetry underlying both is identical.

The first systematic analysis of how to generate all possible sequences
of bells using these simple changes was published in 1668. Richard

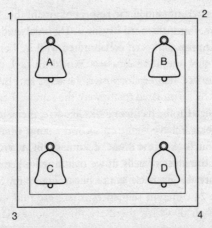

Fig. 68 Using the symmetries of the square to perform Plain Bob Minimus.

Duckworth's *Tintinnalogia: or the art of ringing* explains the principle of how to use basic changes to generate all 120 different permutations of five bells. It could be regarded as one of the first books on groups of symmetries.

Although bell-ringers were exploring symmetry, there were campanology challenges that remained unsolved until the full mathematical language of symmetry was developed. For example, can you get the sequence called Grandsire Triples consisting of $5,040 = 7 \times 6 \times 5 \times 4 \times 3 \times 2 \times 1$ permutations of seven bells by combining so-called triple changes? A triple change is one where three pairs exchange. It took till 1886 for William Henry Thompson, a scholar of Gonville and Caius College, Cambridge, to show that the mathematics of permutations and symmetry implied that this was impossible.

5 April, mudbank on the Thames

The bells of Southwark Cathedral ring out to remind us that time is pressing on. There is another cello piece that Dorothy has brought along which she is particularly keen that I see and hear. It's by a composer who perhaps more than any other has captured the power of mathematics as a canvas for composition.

Iannis Xenakis was a modern Greek composer whose geometric

skills would have commanded the respect of any of the Ancient Greek mathematicians who influenced his ideas. He combined music with a career in engineering, and even collaborated with Le Corbusier on the Philips Pavilion at the 1958 Brussels World Fair. 'I discovered on coming into contact with Le Corbusier,' he said, 'that the problems of architecture, as he formulated them, were the same as I encountered in music.' The design for the pavilion looks like an exercise in Riemannian geometry. Xenakis believed that music was a kind of architecture in sound, and if you look at the score of some of his music, such as that for *Metastasis*, the clustering of notes makes it look remarkably like the diagrams Xenakis drew for the pavilion (Figure 69).

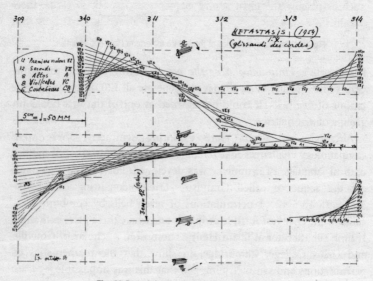

Fig. 69 Part of the score of *Metastasis* by Xenakis.

It was his piece *Nomos Alpha* for solo cello that Dorothy thought would most appeal to my symmetry sensibilities. Xenakis has taken the geometry of the cube and used the symmetries of this object as a framework for his composition. It is a natural extension of what Bach was doing in the *Goldberg Variations*, but Xenakis instead uses the symmetries of the cube. The eight vertices of the cube are used as markers for different sound elements. For example, some corners correspond to dynamics, others to sound textures such as pizzicato and glissando. These sound elements are then played in a particular order according to how the cube is placed on the table. By performing symmetries on the cube, Xenakis alters the order in which these sound elements are applied to the music. The cube is being used in a similar fashion to the way the symmetries of the square helped order the sequence of four bells being rung by the seventeenth-century campanologists.

The cube restricts the possible permutations: because of its rigidity, the eight vertices cannot be arranged in any order. And this is why the structure of the piece will reflect in some mysterious way the rigidity of the cube. If any combination of these eight points were allowed, such as placing the eight sound elements on a pack of cards, there would be $8 \times 7 \times 6 \times 5 \times 4 \times 3 \times 2 \times 1 = 40,320$ different variations. But by forcing the cards to sit on the vertices of the cube, the symmetries of the cube restrict the possibilities to just 48 shuffles.

Interestingly, the cube has the same number of symmetries as the group of symmetries used by Schoenberg and by Dorothy to create their matrix of notes. But the underlying symmetry groups are completely different in structure. The group of symmetries Xenakis is exploiting is called $C_2 \times S_4$, or the direct product of the cyclic group of order 2 with the symmetric group of degree 4. Xenakis extends his idea of symmetries in music in his composition for 98 instruments, *Nomos Gamma*, in which the pyramid and other shapes join the cube in determining the structure of the piece.

Is it important for the artist that people should hear the cube hiding behind the music? One of the interesting things which emerges from my discussion with Dorothy is how a musician or an artist will often deliberately hide the inspiration for a piece. Music is not meant to be proscriptive in its meaning: the artist creates a piece as a catalyst for a multitude of responses to the work. It is a bland piece of art that elicits a uniform reaction. For Dorothy, ambiguity is an important part of art.

What is intriguing is that I have spent the morning trying to take the ambiguity out of my creation. When people read my paper I don't want them to arrive at a completely different conclusion from mine. Ambiguity is anathema to the mathematician. My discovery is like a distant mountain and the proof is the pathway to that summit. I don't want to find readers have arrived at a different summit. Well, actually,

I don't mind if they arrive somewhere else *as well,* just as long as they've visited my mountain too.

Although ambiguity might be something that mathematicians will do their utmost to avoid when writing up their work, they are often as coy as artists when it comes to revealing their inspirations. I appreciate that this is important in the arts, but I believe that it can be quite dishonest in mathematics. I've often read proofs in which suddenly the argument starts to twist and turn in a most surprising manner. Although it all makes logical sense, why the author is tracing this particular path is a mystery. Often the mathematician is keeping some reasoning or helpful picture from view – some sort of secret map which is guiding the direction of the proof.

Gauss was one of the greatest magicians when it came to presenting proofs. Many of his proofs hide a beautiful geometric idea that he had in mind when he made his discovery, but he would keep this picture from view once he came to write it up. If challenged, he would respond that 'an architect does not leave up the scaffolding once he has finished the building'.

The fact that mathematical structures offer such fertile ground for those creating music is one explanation for why the two disciplines have always been considered close cousins. Some people have questioned whether one can really hear the symmetries used by composers, given that a piece of music is linear in time. Unlike the symmetry of a building, where the eye can take in at one glance the balance in the architecture, the fact that music takes you on a linear journey through the composition denies you the chance to hear the symmetry. The musicologist Jean-Claude Risset of the Laboratory of Mechanics and Acoustics in Marseilles wrote that, 'Whereas symmetry is a property of space, time is irreversible. Time's arrow distorts symmetry.'

There is certainly some truth in what Risset says. Time's arrow does insist that a piece be played in a certain direction, and we never hear a piece played backwards. A mathematical proof too is a very linear work, with its own logical arrow of time. Yet the beauty in a piece of music often reveals itself only when you listen to it again. I often need to hear a piece of music many times before it makes any sense to me. I begin to appreciate how themes later on in the piece are echoing or mirroring ideas I heard earlier.

I read mathematical proofs in exactly the same way. My first reading rarely gives me a true understanding of the proof, even though I might follow each of the logical steps along the way. The real understanding comes as I read through the proof again and again. Then I start to see how the author has cleverly manipulated the opening theme – perhaps an equation that I am happy with and understand – and by twisting and turning it, introducing new themes, and interlacing these themes, has transformed the proof into something surprising, leaving me with a sense of having journeyed into unfamiliar territory.

The physical response that music can stimulate in a listener's body is perhaps harder to achieve in mathematics. The hairs on the back of my neck always stand on end at the same moment in the recording I have of Richard Strauss's Four Last Songs when, in *Frühling*, Lucia Popp's voice swoops down and then up again. Mathematics does give me an emotional response, but perhaps not quite to the same degree as music. That rush of emotion is an essential part of recognizing when I've made a mathematical breakthrough. As the brain recognizes a new pattern, the body releases a shot of dopamine to make sure I don't miss it.

The Thames tide is coming in by the time we've finished exploring Xenakis and the symmetries of the cube. We'll be washed away if we don't move. Dorothy is going to read a book I've recommended which explains the mathematical language of symmetry: a Group Theory Primer. I've got the easy job of listening to music.

May: Exploitation

Of what use is symmetry?

MAO TSE-TUNG

4 May, Oxford

One of my graduate students has just left my office. He's done some great work over the past three years and is starting to write up his doctorate, but he's just confessed that he's not sure that he wants to be a mathematician. I'm feeling quite sobered by this news. My graduate students are like my children. They are the future of the subject. Who's going to read all the details of my papers if not my mathematical offspring? The subject feels so tribal that anyone who says they want out is almost a threat to everything the tribe stands for.

Anton has been working on a project very close to my current problem. There's no denying that one can feel quite disillusioned by not finding a way into a problem. Last year one of my post-docs left for the City after attempting to scale this mountain with me. I'd already rescued him from being dragged off to the City once before. But after battling with our problem and seeing it become more and more complex, he felt that he wasn't really cut out for it.

What is unsettling for me is that they both questioned the importance of what we are doing. They've asked that 'What's it all for?' question, and think they've seen the Emperor without any clothes. Anton has questioned whether the problems we are working on are really important. I've explained why I think these are fundamental questions about basic objects in nature, but I can see that he isn't convinced. I feel I am having to defend my whole existence. I've arranged for him to join me at a conference in Israel later this month, and I hope that seeing the rest of the tribe enthused and excited about these problems will re-inspire him. It will also show him that people are interested in what he is dedicating his time to.

Sometimes the very abstract and unworldly nature of the subject can get you down. You dedicate years of your life to cracking a conjecture, and when it's done there are only a few people in the world who will be able to appreciate it. Your family and friends haven't really got a clue what it is you've been up to. You can try to give them a feel for the thrill of the breakthrough, but sometimes you just wonder what the point of it was when the audience is so small. I often envy scientists in other fields for the immediate respect they command from onlookers.

I visited a laboratory a few years ago. The lab assistant put a Petri dish under the microscope and let me look at the contents. It was a fertilized human egg that had multiplied into four cells on its way to becoming a human being. It was staggering. Understanding what was happening in that dish seemed like real science. It mattered, it felt important. Whether my conjectures about symmetry were true or not seemed to pale into insignificance as I peered into the microscope. But it was also the future of our family that I was looking at, which certainly accentuated the relevance of this particular piece of science.

Unfortunately, the science didn't work in our case. The four cells didn't make it up to the 26 billion cells that constitute a new-born baby. I chose maths over the other sciences because, if a proof works once, it always works, every time it is repeated. Physical experiments, on the other hand, often go wrong. I can't really cope with the uncertainty and lack of control in the physical world, which is why I'm drawn to the clean, unforgiving logic of the mathematician's lab.

When I was younger I used to revel in the unworldly nature of my subject. I was under the spell of G. H. Hardy's *A Mathematician's Apology*, in which he lays out a manifesto for why mathematics should be celebrated for its own sake and why we shouldn't be driven by the desire to see our work applied. But as I've grown older I've changed my tune a bit. I still study very abstract problems, but I would love it if something I discovered suddenly found a practical application. And it is not unreasonable that this might happen.

A huge amount of mathematics goes into understanding the science that I was looking at through the microscope in that laboratory. As the cells double up each time, symmetry is playing a crucial role in determining the possible configurations for the new cells and ultimately for the shape of our bodies. Although I might not directly be trying to apply the mathematics I discover in the mathematical laboratory of my room in Stoke Newington, nevertheless, these breakthroughs form part of a chain of unexpected connections which leads ultimately to the extraordinary uses of mathematics in our everyday lives. And because of this chain, something which at first sight looks terribly abstract can be precisely the piece that is missing from a jigsaw, something that can explain the mystery of life. Indeed, it was the power of the microscope that during the twentieth century revealed that symmetry underlies many of science's greatest mysteries.

Tasty tetrahedrons and poisonous pyramids

With the development of more sophisticated microscopes towards the end of the nineteenth century, scientists suddenly had access to the small-scale structure of matter and they discovered a whole new arena of symmetry at work in the natural world. The building blocks of crystals and gemstones, tissues and bone, and cells and viruses all exploit a variety of the symmetries that are possible in three-dimensional space.

One of the shapes that chemists found pervading the molecular world is the tetrahedron – the triangular-based pyramid built from four perfectly symmetrical equilateral triangles. The fact that carbon, the chemical of life, often connects to four other atoms means that carbon-based molecules are often tetrahedral in shape. The carbon atom sits at the centre of the shape and the four other atoms are located at the points of the pyramid.

The most common example of such molecules is methane, which consists of one carbon atom and four hydrogen atoms arranged around the central carbon atom (Figure 70). Each hydrogen atom positions itself so that it is as far away as possible from the other three hydrogen atoms. Like a stone falling to the bottom of a hill to a position where its energy is minimized, the four hydrogen atoms seek the vertices of

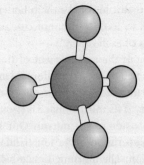

Fig. 70 The tetrahedral configuration of the four hydrogen atoms in a molecule of methane.

the tetrahedron as the arrangement of lowest energy. The tetrahedron is to methane what the sphere is for a bubble. Symmetry provides nature with an arrangement which minimizes energy.

Chemists enjoy representing molecules as models made from different coloured ping-pong balls and sticks. In the methane molecule, the hydrogen atoms are represented by four balls of the same colour placed around a larger ball, the carbon atom. Although other carbon-based molecules can be very much more complicated in structure, they can often be regarded as essentially tetrahedral in shape. Chemists still like to represent these by models where four different coloured balls, each signifying a complicated molecular structure, are arranged in a tetrahedron around the central carbon base. In the 1950s the pharmaceutical company Grünenthal found that one such tetrahedral molecule produced a drug that stopped morning sickness in pregnant women. They marketed the drug under the name thalidomide. It was a popular product – until the frightening discovery that mothers who had taken it were giving birth to children with deformities such as very stunted limbs.

It turned out that although the tetrahedral molecule built by Grünenthal was perfectly safe and did suppress morning sickness, it had a symmetrical cousin that was very dangerous. There are in fact two different ways to arrange the four coloured balls around the carbon atom. There's red on top, and then, below the red going clockwise, blue, yellow and green, and there's also blue followed by green then yellow. These two arrangements are distinct molecules. There is no way to rotate one so that it looks like the other. The only way to turn one into the other is to look at the molecule in a mirror. The two molecules are mirrors of each other.

It is the asymmetry in the arrangement of the points of the tetrahedron that produces two different thalidomide molecules. Other pairs of otherwise identical molecules also have this relationship. A molecule is said to have chirality if its mirror image is genuinely different to the molecule on the other side of the mirror. Our hands have chirality: there is no way to superimpose the left hand on the right – they match only in reflection. The morning sickness drug turned out to be unstable, so that even when Grünenthal had synthesized one arrangement of the molecules, half of them would degrade into their mirror image, producing the other dangerous arrangement. So depending on how you build your tetrahedron, you can end up producing a poisonous pyramid rather than a therapeutic tetrahedron.

Our body is extremely well tuned to noticing the difference between mirror versions of these tiny pyramids. The smell of one arrangement can be totally different to that of its mirror image. For example, one version of the chemical carvone smells of caraway, while its mirror cousin smells of spearmint – and is smeared on Wrigley's gum to give it its characteristic taste. It is quite extraordinary that our senses can sniff and taste symmetry. In *Through the Looking Glass*, published in 1899, Alice suggests that milk through the looking glass may not be so good for her kitten to taste as ordinary milk. Lewis Carroll was well ahead of his time.

The reason that our bodies react differently to symmetrical cousins of the same molecules is that the amino acids found in proteins in the body always seem to be pyramids that are arranged with one particular chirality – you never see its mirror image. Scientists have called these left-handed pyramids. A right-handed version of a drug might react therapeutically with the proteins in our body while its mirror cousin combines destructively with the same protein, a more dramatic version of the way shaking someone's left hand feels awkward.

It seems that living organisms are always made of left-handed pyramids. Why this should be so is still something of a mystery. Is this propensity for left-handed pyramids true of all organic matter across the universe? Or is it just our corner of our galaxy that favours one version of the tetrahedron over the other? Scientists talk about the left-handed universe, but no one really knows which wind is blowing to cause our amino acids to spin that way.

The connections between symmetry and molecular structure led during the course of the twentieth century to a new dialogue between chemists, biologists and mathematicians. The other sciences began to tap into the power of mathematics to reveal the different shapes that were possible for these molecular configurations. Indeed, microbiologists discovered that some of the other Platonic symmetrical shapes were at the heart of one of the most dangerous forces of nature: the virus.

Viruses: why symmetry makes you sneeze

At the end of the nineteenth century, tobacco plants in the Crimea were being destroyed by an unknown cause. A young Russian biologist from the University of St Petersburg, Dmitri Ivanovski, was dispatched to try to identify what was causing the 'wildfire' that seemed to be afflicting the plants. It was assumed that a bacterial agent was responsible, but when the biologist analysed the agent that seemed to be causing the disease, it seemed to be very different to bacteria. Bacteria cannot permeate a porcelain filter, but the minuscule entities that seemed to be destroying the plants passed clean through. It was a Dutch microbiologist, Martinus Beijerinck, who named these new infectious particles 'viruses' and guessed that they were in some sense living organisms that used the cells of plants or animals for reproduction.

It was Spanish Flu that really concentrated the mind on the science of viruses. In 1918 the Spanish Flu pandemic killed in the order of 50 million people – more than the casualties of the First World War. Suddenly, scientists were very keen to understand the mechanism of this dangerous disease. The structure of bacteria could be seen under a conventional microscope, but an influenza virus was too small. It would need the more subtle techniques of X-ray crystallography and the invention of the electron microscope before the make-up of the tiny organism could be penetrated. Francis Crick and James Watson, after they cracked the structure of DNA, were two of the scientists who turned their attention to the nature of viruses.

Rather than a tangled mess, scientists found instead an object filled with symmetry – albeit a deadly one. As it happened, many artists at the beginning of the twentieth century had begun to associate symmetry with death. In *The Magic Mountain*, Thomas Mann's protagonist contemplated the symmetry in a snowflake and 'shuddered at its perfect precision, found it deathly, the very marrow of death'. Now symmetry was a symbol of death not only for the artist but also in biology.

Since the 1930s it had been known that these viruses consisted of a piece of genetic material called RNA surrounded by a shell of protein. From the images of the virus that X-ray crystallography was providing, it was apparent to Watson that the shell of the virus that attacked and destroyed tobacco plants looked like a spiral, whereas the images of other viruses, such as the Tomato Bushy Stunt Virus, suggested a more spherical shape. Watson discussed his ideas with a young research student, Donald Caspar, during a meeting at Cold Spring Harbor in summer 1954. Caspar came to England where he teamed up with a young post-doc, Aaron Klug. Caspar worked on the Tomato Bushy Stunt Virus, and Klug on Turnip Yellow Mosaic Virus. To study the shapes of such small viruses, they passed X-rays through a crystallized version of the virus and then analysed the resulting diffraction pattern. From these images it appeared that both viruses were demonstrating a more spherical shape, but Caspar and Klug were keen to pin down the precise design of these infectious agents.

The process of reconstructing the shape of a crystal from the images produced by X-ray crystallography is rather like doing a complex geometric puzzle in three dimensions. Imagine that someone has arranged a lot of balls in some configuration. Your challenge is to discover just how these balls are arranged. However, the only information you are given is a handful of two-dimensional shadows of the configuration. Essentially, the shadow-patterns look as though someone has stamped on the three-dimensional structure and squashed it onto the two-dimensional page. You have to give life to the dots in the picture and describe the three-dimensional arrangement that was there before.

The diffraction patterns that Caspar and Klug obtained seemed to indicate that there were four axes about which the crystal had rotational symmetry of order 3, like a triangle. When they looked for mathematical shapes that have these symmetries, they found that all the Platonic solids have four such axes. The axes of a tetrahedron, for example, are easily identified: poke a stick through one vertex and pull it out through the centre of the opposite triangular face. You can spin the tetrahedron through a third of a turn round this axis, and the shape looks unchanged from this new angle.

Intriguingly, the cube also has a rotational symmetry of order 3, like a triangle. Why should this be? The thing is made up of squares, so where are the triangles? If you take a stick and push it through one of the cube's vertices and then pull it out through the vertex farthest from it, you can spin the cube around the stick as an axis. Rotate it through one-third of a turn, and the cube will look the same. You can see the triangle if you look at the edges emanating from the vertex: three edges meet at each vertex, and they realign after each third of a rotation. Another way of seeing a triangle in a cube is to cut off one corner – a triangular face will then be staring at you.

The existence of these four rotational symmetries of order 3 seemed to be pointing Caspar and Klug towards one of the five Platonic solids as a potential shape for these viruses. But which one? The more views and different perspectives you get of the arrangement via this technique of X-ray crystallography, the better your chance of working out what the shape looks like. Caspar struck lucky and obtained an image of the Tomato Bushy Stunt Virus which showed five dots in a pentagonal configuration. This two-dimensional picture indicated that the proteins in the shell of the virus were arranged in such a way that there was an axis with a fivefold rotation like that of a pentagon. This narrowed down the Platonic solid on which the virus might be modelling itself. Of the five Platonic solids, only the 20-sided icosahedron and the 12-sided dodecahedron have rotations like that of a pentagon. Caspar showed the pictures to Crick, who was working in the same laboratory in Cambridge.

Crick and Watson started to formulate a theory of the mechanics of how the proteins are arranged in the virus's shell. The way the virus reproduces itself is to inject the RNA at its heart into a host cell and then sneakily use the cell's machinery to replicate itself. Encoded in the RNA is the information to build and assemble the proteins that will surround the replicated viruses created by the host cell. Once assembled and packaged in the newly built protein shell, the cell then releases these new virus particles to wreak havoc on more host cells.

What makes a virus so harmful is the effect it has on the host cell it leaves behind in the process of replicating itself.

The experimental evidence indicated that the string of RNA released by these viruses was very short and would therefore be limited in the amount of information it could encode. Some of the viral genomes are so small they contain fewer than five genes. Basically one can think of the released RNA string as a little computer program which is used to build a protein element and then assemble the elements in a particular configuration. Contained in human DNA, for example, are extremely complicated programs which are used to build the heart and other organs in a human foetus. The same mechanism was responsible for the cell division I'd marvelled at under the microscope that day in the IVF lab. What Crick recognized was that the small length of the program contained inside a virus meant that a new structure would necessarily have to be built in a very simple manner. And this is the power of symmetry.

The symmetry of an object essentially provides a very simple program for constructing the whole of the object from a simple building block. The helical shape of the Tobacco Mosaic Virus, for example, is very simply put together: the helix looks like a spiral staircase in which each step is an identical piece of protein (Figure 71). Each full turn of the staircase has $16\frac{1}{2}$ steps. The same rule is applied at each step:

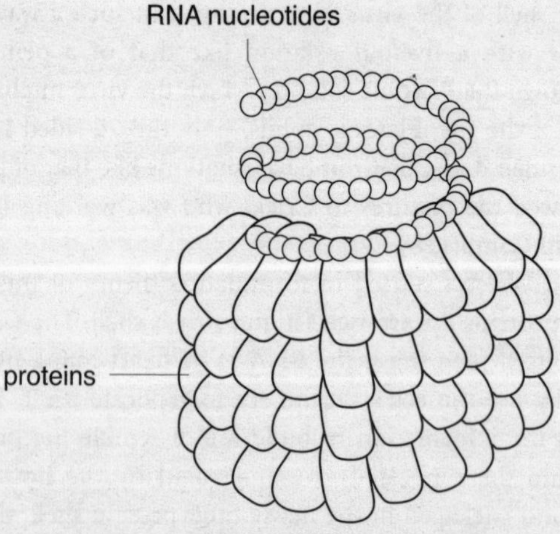

Fig. 71 The helical shape of the Tobacco Mosaic Virus.

simply make the appropriate twist of the staircase and add the next identical piece of protein. What about the viruses with the four rotational axes through them – what rule could apply for them? The beauty of the Platonic solids is that any two faces meet each other in exactly the same way. There is no part of the shape which requires an extra rule to construct it. So again, the rule for putting these shapes together is as simple as for the staircase.

Crick believed that the underlying shape of some of the more spherical viruses was a 20-faced icosahedron in which each triangular face was made up of three protein elements, making 60 protein elements in total in the shell (Figure 72). The underlying symmetry immediately helps to reduce the size of the program required to reproduce the structure. This is one reason that nature is attracted to symmetry, as a labour- and information-saving device. It's also how Mozart was able to remember Allegri's *Miserere*.

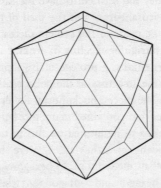

Fig. 72 Some viruses are icosahedral in shape and consist of 60 identical protein pieces.

The idea of assessing the complexity of a structure by using the length of a program to generate it has been used recently in number theory. The program required to reproduce, say, the infinite decimal expansion of $1/3$ is very short: just keep repeating 3. The program for generating the decimal expansion of π, on the other hand, is much more complicated. This is why you get in the record books for remembering the expansion of π but not for remembering the expansion of $1/3$. There is a whole branch of mathematics dedicated to establishing the length of the programs required to reproduce certain numbers.

The more symmetry a number or an object has, the shorter the program needed to reproduce it. Conversely, Crick believed that the concise program encoded into the short string of RNA at the heart of a virus would necessarily result in a structure with symmetry.

Crick presented his ideas on the shape of spherical viruses at a meeting of virologists, but he was met with some scepticism. However, Caspar had come armed with models made from sticks and ping-pong balls to show the possible configurations. As the structure took shape in front of them the virologists became more convinced by Crick's vision. But they couldn't see how the model would explain more complex viruses that were coming to light. The problem was that if a virus was so symmetrical, then the geometry allowed for a maximum of 60 protein pieces. It appeared, though, that viruses sometimes contained a lot more than 60 pieces in the shells that encased the RNA. Klug and colleagues had started studying animal viruses in addition to plant viruses, and they too seemed to have the same rotational symmetry found in the icosahedron. But the shell of the polio virus, for example, seemed to be made of 180 protein pieces.

At this point, an intriguing meeting of art and science pointed the way forward. Robert Marks, a protégé of Buckminster Fuller, came across the work on the structure of the polio virus which seemed to be constructed from 180 pieces. Buckminster Fuller had been building domes from triangles that contained many more than the 60 that the mathematicians seemed to be limited to. But in these domes not all the triangles were identical – they nearly were, but not quite. For example, take a football made up of 20 hexagons and 12 pentagons. Now subdivide each hexagon and pentagon into triangles. The triangles in the hexagon are all equilateral and perfectly symmetrical. But the triangles in the pentagons are not quite equilateral. Now I have a structure with 180 faces built out of two distinct building blocks (Figure 73). So perhaps the more complex structures such as the polio viruses looked like a triangulated football, or one of the domes of Buckminster Fuller.

Klug met with Buckminster Fuller in the summer of 1959, but it was reading Marks's book published in 1960 on the life and work of the architect that led Caspar and Klug to reformulate the structure of the virus. Although their new structure was still full of symmetry, they introduced the idea of quasi-equivalence – a slight relaxation of the

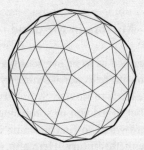

Fig. 73 A polyhedron made up of 120 equilateral triangles and 60 isosceles triangles.

constraints on the relationship between the pieces in the shells. So just as the geodesic dome is made from piecing together two slightly differently shaped triangles, it could be that two slightly differently shaped protein pieces combined to form the polio virus. There were still constraints on the number of pieces, and Caspar and Klug came up with a formula to calculate the possibilities, and even borrowed some of Buckminster Fuller's architectural terminology. La Géode, which Tomer and I visited in Paris, is an example of how all these different triangles can be pieced together to approximate the sphere, the shape with ultimate symmetry.

Klug and Caspar also came up with a different model for the formation of the shell of the polio virus. Again, symmetry came to the rescue as a way to collect the 180 protein pieces together. Just as the perfectly symmetrical sphere is the surface of minimum energy which a soap bubble aims for, the same trick works for the shell of the virus. The most symmetrical way to assemble the 180 pieces represents the minimum energy and therefore the state of rest for the virus shell. Symmetry helps the process of virus construction.

Recent studies have revealed that some of the most deadly and virulent viruses have the icosahedron as their shape of choice: herpes, rubella, even the HIV virus which causes AIDS – all hide their deadly secrets within a highly symmetrical skin. But modern science has revealed that symmetry is key not only to the very tiny organisms of nature, but to one of the greatest mysteries of biology – the functioning of the mind itself.

Mirrors in the mind

Evolution has programmed us to be oversensitive to symmetry. Those that can spot a pattern with reflectional symmetry in the chaotic tangle of the jungle are more likely to survive. Symmetry in the undergrowth is either someone about to eat you or something you could eat. Our brains seem to be hard-wired to find meaning in symmetry. This is why, in the early twentieth century, Hermann Rorschach developed his symmetrical inkblots as a means of unlocking a patient's unconscious mind. He believed that humans are so compelled to find a meaning or a message when shown something symmetrical that the patient's response can reveal clues to their psychological state of mind. Carl Jung also thought that symmetry was important in understanding the unconscious. But rather than the butterfly-like inkblots of Rorschach, Jung was drawn to the symbolism contained in the mandalas of Hinduism and Buddhism.

In Buddhist worship the mandala is a sacred diagram made from dyed sawdust or sand. *Mandala* is Sanskrit for circle, and the diagram is an intricate network of intertwining circles representing a many-layered universe. The complexities of the design are meant to reflect the turmoil of the human condition on the way to achieving the nirvana at the centre of the mandala. Each symbol in the mandala's design has some specific meaning. A diamond represents the mind, the eight-pointed wheel denotes the eightfold path to Nirvana and the 16-leafed lotus is Buddha or Nirvana itself. The network of shapes acts as a story, guiding the worshipper in a lesson on meditation.

The mandalas of Tibet are laboured over by priests for days in an act of deep and concentrated worship. The symmetry of the shape is meant to assist the worshipper in achieving a meditative state, yet once finished they are immediately destroyed – an important part of the mandala ritual. The hands that have painstakingly worked for perfect symmetry pass over the sand, and the symmetry is lost. It is a lesson on the impermanence of the world and the fragile nature of human existence. Symmetry is hard to achieve, as those in the animal world with bad genes discover to their cost. The destruction of the mandala also reinforces the Buddhist belief in the importance of non-attachment. Nirvana comes only to those who are prepared to let go of the things they are most attached to. A Buddhist monk who has spent days building a beautiful picture will tell you that the destruction of his creation, full of delicate symmetry, is much harder than the hours spent making it. Once destroyed, the coloured sand is thrown into moving water.

For Jung the mandala was an expression of the self:

> I sketched every morning in the notebook a small circular drawing, a Mandala, which seemed to correspond to my inner situation at the time. With the help of these drawings I could observe my psychic transformations from day to day.

Jung also used his patients' drawings of mandalas as a doorway into their subconscious world. He believed that the act of creating these symmetrical images was also therapeutic in its own right and helped patients to express the different facets of their personalities:

> Most mandalas have an intuitive, irrational character and, through their symbolical content, exert a retroactive influence on the unconscious. They therefore possess a 'magical' significance, like icons, whose possible efficacy was never consciously felt by the patient.

Jung was very struck that when asked to express the inner turmoil of the mind through drawing, different patients would often sketch the same basic shapes. The fact that the same imagery was thrown up by so many different patients and across so many diverse cultures supported Jung's belief in the idea of a collective subconscious:

> There must be a transconscious disposition in every individual which is able to produce the same or very similar symbols at all times and in all places. Since this disposition is usually not a conscious possession of the individual I have called it the collective unconscious, and, as the basis of its symbolical products, I postulate the existence of primordial images, the archetypes.

For Jung, the triangle, the circle and the hexagon were all universal symbols that resonate with the human condition regardless of cultural background.

Psychologists have developed another test which highlights our sub-conscious attraction to symmetry. Try this exercise. You are shown the four cards in Figure 74. Each card has a letter on one side and a number on the other. You are told that every card with a vowel on one side has an even number on the opposite side. Which card or cards must you turn over to check whether this statement is true?

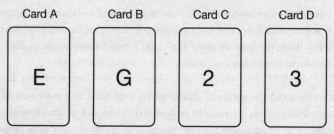

Fig. 74 Which card or cards must you turn over to check whether every card with a vowel on one side has an even number on the opposite side?

Psychologists talk about two modes of reasoning and memory: the old mode and the new mode. The old mode corresponds to our most basic primordial animal brain, and it seems that this is the one that wants to find symmetry everywhere. We tend to use this mode in our intuitive response to problems. In experiments, only 4 per cent of subjects were found to have the ability to override their old brain pathways and apply a more analytic response to get this puzzle right. Most people think that they must check card A and card C. But actually you need to check card A and card D. Why is this? You have to check whether every card with a vowel on one side has an even number on the other. Our brains are so desperate for symmetry that they also check whether every card with an even number on one side has a vowel on the other. So the brain directs you to pick up card C. But if that card has a consonant on the other side, so what? It's card D that might destroy the theory. If card D has a vowel on the other side, then the theory is false.

It seems that our subconscious works according to a sort of sym-metrical logic. It thinks that if the statement 'if A then B' is true, then so is its converse, the mirror image 'if B then A'. In general, this is far from true. Logical deduction is usually very unsymmetrical. There is a whole school of psychology that tries to explain the mechanism of the

subconscious in terms of this desire for a symmetrical logic. The father
of this movement is a Chilean psychologist, Ignacio Matte Blanco. His
model of the subconscious looks something like the edifice Borges
describes in his short story 'The Library of Babel'. According to the
narrator, every possible book is meant to exist somewhere in the
library. So there are geography books in which Paris is the capital of
France, but also a book in which France is the capital of Paris. Borges'
physical description of the library as a lattice of interconnected hexagonal cells sounds like the description of a huge brain. Matte Blanco
believes that the human mind, like the library, makes these strange
identifications which arise from a desire to balance things.

The propensity of the brain to home in on symmetry can have
devastating effects on a psychotic patient. Matte Blanco described a
schizophrenic patient who became terrified after an incident when a
sample of blood was taken from her arm. At times she would say that
blood had been taken from the arm, but at other times she would
completely reverse the statement and declare that her arm had been
taken and the blood left behind.

Matte Blanco believed that, like the books in Borges' library, the
patient's brain has used symmetry to allow for all permutations of a
statement. He extended the ideas of Sigmund Freud, whose theories
also suggested that our brain attempts to ascribe symmetry where there
isn't any. 'My father is a man,' says our conscious brain, which in our
subconscious becomes 'All men are my father.' These symmetrical
games are particularly powerful in our dream world, where the subconscious takes over. The twisted world that this inner symmetrical
logic produces explains the bizarre turns that our dreams can sometimes take. Perhaps subconsciously I think that a friend of mine is
actually rather dangerous. My conscious agreement with the statement
'Monsters are dangerous' can suddenly, via this symmetrical logic that
operates in the dream world, turn my friend into a monster.

It is beginning to emerge that symmetry plays a fundamental role
in how the brain works. The discovery of special neurons called mirror
neurons is regarded as one of the great breakthroughs in neuro-
physiology. Like many great scientific breakthroughs, it came about by
chance. Three scientists working at the University of Parma, Giacomo
Rizzolatti, Leonardo Fogassi and Vittorio Gallese, were exploring which
neurons in the brain fire when monkeys move their hands in certain

ways. These neurons are called motor neurons because they deal with motor skills. The scientists attached electrodes to the frontal cortex of the monkeys. They were able to identify specific neurons which would fire for each particular motion. When the electrodes were wired to one particular place in the brain, each time the monkey reached for a peanut the machine would emit a 'ssshhhh' sound to indicate that the neurons were firing.

Leonardo Fogassi had spent the day watching the actions of the monkeys and recording the corresponding 'ssshhhh' sounds that the machine emitted. Happy with the day's findings, he began to tidy up the lab. But as he reached out to collect up peanuts that were lying around, the machine suddenly went 'ssshhhh'. 'That's weird,' he thought. As he reached for another peanut, he saw the monkey following his hand movement with its eyes – and there was another 'ssshhhh'. But the monkey hadn't moved its hand at all. Perhaps the equipment was faulty. Would the whole day's research have to be scrapped? Fogassi checked the equipment, but there appeared to be nothing wrong.

Although the monkey was not actually taking a peanut, it seemed that neurons in its brain were firing as if to create a virtual reality version of the action. These weren't motor neurons firing, but something that the researchers christened mirror neurons or 'monkey do, monkey see neurons'. The idea that neurons fired to imitate or mirror the act of another animal could provide vital clues to the development of the human mind. Indeed, the neuroscientist Vilayanur Ramachandran has predicted that mirror neurons will do for psychology what DNA did for biology.

Seeing someone else do something looks very different from the image you have of yourself performing the action, yet the brain activity is almost identical. Something in the brain seems to suppress it from sending a signal to actually perform the action. But sometimes this doesn't work, and a neuron fires and your body copies the action you've seen. How many times have you caught yourself mirroring the movements of someone you're talking to? It cracks me up whenever I notice that I've just put my hands behind my head in exactly the same way as the person opposite me. Mirror neurons are firing and producing this desire in my body to mirror in perfect symmetry the action of the person in front of me. If somebody yawns loudly in front of you, it's not long before you are yawning too, even if you aren't the least bit tired.

The mirror neurons help to explain the striking ability of babies to simulate so perfectly the facial movements of their parents, even though they haven't a clue what their own faces look like. When its parents stick out their tongue, a baby is able to repeat the action. The baby doesn't need hours of practice in front of a mirror to copy the facial expression; instead the mirror neurons fire, making a copy of the action in the baby's brain. These mirror neurons may have helped humans to develop their sophisticated language skills. The acquisition of language depends on mirroring the sounds of others. Like a frieze pattern round the top of a pot, the child hears a sound made by the parent and tries to produce a perfect copy of it. Researchers have indeed found that the part of the brain where the mirror neurons are located is similar in position, structure and evolutionary origin to the Broca's area of the brain which deals with language.

Some scientists have suggested that something triggered an explosion of mirror neurons in the human brain around 40,000 years ago. That was when there was a Big Bang in human cultural development. Suddenly, tools are becoming more refined. Lumps of rock are being carved into symmetrical arrowheads. Tools are covered in interesting patterns, not for any utilitarian purpose but because the brain is becoming increasingly attracted to symmetrical forms during this period.

The ability of these mirror neurons to help us to get inside other people's heads is regarded as the key to binding humans together in groups with clear cultural identities. Humans have been referred to as 'the Machiavellian primate' because of this ability we have to comprehend the actions of others. Indeed, mirror neurons might hold the key to understanding autism. A failure to empathize might be due to the failure of these mirror neurons to fire when someone is observing other people. Empathizing requires this virtual reality machine inside our heads to simulate what it feels like to be doing what we see others doing.

Mirror neurons could well be behind our very sophisticated means of communication by language and speech. And in the twentieth century, symmetry was exploited to facilitate an explosion of electronic communication that has swept the globe.

You're cracking up . . .

Symmetry has become the key to the survival of data as it flies through the busy airwaves. The integrity of a message depends on mathematical codes that exploit the mathematics of symmetry. The word 'code' usually conjures up images of spies and secret messages. But the technical term for a system that scrambles a message, like the German Enigma machines of the Second World War, is a cipher. The origin of the word comes from the Arabic for zero: *sifr*. In medieval Europe, the new numerals from the East – including zero – gave ordinary people increased access to the power of computation. In order to maintain their superior position, the upper classes banned the use of these numerals, forcing people to use them in secret. This is how the word for zero came to mean a system for keeping messages secret.

The technical meaning of 'code' is actually a system that preserves a message and aids its communication – almost the opposite of a cipher. Life itself is transmitted and reconstructed through code. DNA is just a long sequence of four molecular compounds: adenine, guanine, cytosine and thymine. These are generally represented by the letters A, G, C and T. The DNA molecule can be represented as a long code word consisting of these four letters. The string of symbols encodes information about the parents' genes and information to reproduce another human being. In a virus, the RNA encodes how to rebuild another copy of the virus out of host material.

A lot of research has gone into how good this code is in correcting errors, which would produce a mutation – a change in the genetic make-up of an organism. The individuality of each human being depends on some of these mutations occurring: none of us is an exact clone of a parent. However, there needs to be a mechanism in place to correct major errors in the transmission of the DNA from parent to child, to prevent the program for reconstructing life from crashing.

One code which is particularly effective at retaining meaning, even in the face of multiple errors, is the written word. A tetx wiht srcaabedl ro chnyade ellters osmheow rewains emghuo infwrtmation to neabke the oirginl sendnence ti be reoncsrcted. The written word is so good at correcting itself because it has a lot of redundancy in it. Every word in a dictionary is like an admissible code word. For example, at the time of writing there are 48,504 seven-letter words in the *Oxford English Dictionary Online*, but over four billion possible combinations of seven letters. The extreme unlikelihood of a random string of seven letters being a word in the dictionary makes the written language good at correcting itself. For just a slightly garbled seven-letter word there aren't going to be many alternative choices.

In contrast to written text, the spoken word in general isn't so effective, which is why the game of Chinese whispers is fun. A story tells how the message 'Send reinforcements. We're going to advance,' sent from the front line during the First World War, mutated via word of mouth into 'Send three and fourpence. We're going to a dance' by the time it reached headquarters.

In some sense sudoku, the number puzzle that has swept the world, is a sophisticated error-correcting puzzle. A standard sudoku puzzle consists of a 9×9 grid in which only a few squares have a number in them. You are told that the numbers from 1 to 9 occur once in each row, once in each column and once in each 3×3 block. Because of this internal structure in the table, it is possible to work out all the missing numbers from the ones already there. So a sudoku is like a message in which lots of the information has been lost in transit but, thanks to the structure of the message, it is possible to reconstruct the whole message from the partial signal received.

The explosion in communication in the twentieth century created a never-ending need for cleverer, faster and more efficient ways to preserve messages. The airwaves are filled with mobile phone conversations, digital radio signals and satellite transmissions. CDs, DVDs and MP3 files all store information in digital format that can be reconstructed into the music or video we want to watch. Modems and fibre optic cables are transmitting emails and webpages from one computer to another. We depend on orbiting satellites to collect and send data about the weather, and spacecraft sent to the far reaches of the solar system beam back images of distant worlds.

With so much electronic fluctuation and strong magnetic fields interfering with the digital data, whether travelling through the atmosphere, along cables or through the vacuum of space, scientists have been forced to come up with ways to tell whether data has been corrupted. Is there a way to send data so that even if the message has been damaged in transit we can find a way to piece together the uncorrupted message? A typical CD has over half a million errors in the digital data encoded onto it when it comes off the production line. Once the CD has been used and abused, picking up scratches and fingerprints, those errors can go up to a million or more. This is a small percentage of the overall data on the CD but still potentially disruptive to the reconstruction of the music. The power of mathematics has ensured that the way the music is encoded enables the majority of errors not only to be detected but also corrected. The mathematical journey to such powerful codes starts with finding a way to translate a picture, a voice or a piece of text into a simple string of numbers.

Because computers are essentially systems of switches that are either on or off, the numbers they like best are 0's and 1's: a 0 sets a switch in the 'off' position, and a 1 turns it to the 'on' position. One of the first translations from letters into numbers was made by a French engineer, Émile Baudot, in 1874, for transmitting messages by telegraphy. Each letter of the alphabet became a string of five 0's and 1's. This allowed Baudot to represent a total of $2 \times 2 \times 2 \times 2 \times 2 = 32$ different characters. The letter X, for example, was represented by the string 10111, while the letter Y was 10101.

Quite often a text might require more than 32 symbols, for example symbols for punctuation and numerals, so Baudot came up with a cunning way to expand the range. Just as a keyboard uses a shift key to get access to a whole range of other symbols using the same keys, Baudot used one of the string of five 0's and 1's to denote pressing the shift key. So if you see the string 11011, you know that the string immediately following is from the alternative set of characters.

Although it has been superseded by superior codes, such as ASCII, Baudot's code got a recent outing on an album cover. A newspaper phoned me up last year to ask if I could write an article about a mysterious puzzle that Coldplay had embedded in their recent album. The phone call came through at midday. They needed the article by three o'clock for the next day's paper. I went into a bit of a panic on the phone – it was like my final examinations at university all over again. 'What if I can't crack it by three?' I asked. 'Oh, that will also be interesting,' replied the reporter, 'Oxford professor fails to crack Coldplay cover.' The thought of that headline soon concentrated the mind.

The album cover shows coloured blocks arranged in a grid. I soon guessed that the colours are irrelevant, and that you have to read a block of colour as a 1 and a gap between colours as a 0. So in fact I just needed to take a black and white image of the cover, where black denotes the presence of a colour in the grid (Figure 75). Reading down the first column of the grid, you get black, white, black, black, black, which translates into 10111 – the Baudot code for X. The last column translates into the Baudot code for Y. What about the two columns in between? The second of them gives 11011, the Baudot shift key. So to interpret the third column, 00011, I needed to consult the alternative character set that Baudot encoded. Now I knew that what I was after was the symbol for &. It was no secret that Coldplay's new album was called *X&Y*. But when I looked up the code it seemed that 00011 was the symbol for 9, not &. The cover should have depicted 01011 in the third column. According to its cover, Coldplay's album is called *X9Y*, not *X&Y*.

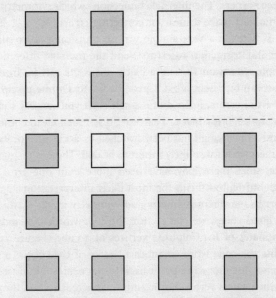

Fig. 75 A black and white version of the code on Coldplay's album cover.

The cover demonstrates one of the problems with Baudot's code. Any interference and you get another perfectly acceptable message.

Obviously there was some noise on the line when the band was speaking to the graphic designer, who changed 01011 to 00011. Because of the nature of the code it was impossible to detect the error unless you knew what the title of the album was. The cover still looks striking, even with the mistake. But it's not only the title of the album that has been converted into a string of 0's and 1's. The CD inside has stored the music in digital format too. Here the ability to correct mistakes, such as errors in printing or scratches, is far more important.

Mathematicians discovered that there were ways to use symmetry to detect errors that had crept into a message during transmission. Symmetry provides the means to detect corruption. The weaver of a carpet whose pattern is full of symmetry can detect and correct minor errors. If the four corners of the carpet are copies of each other yet one of the four is turning out slightly different, the weaver can check and put it right from the information contained in the pattern in the other three corners. The internal connections which symmetry sets up within an object make it ideal for correcting errors.

There is actually a very simple way to transmit data to allow for detection and correction of errors: send the message three times. If, for example, you want to send a black and white picture from outer space, you can represent a black pixel by 0 and a white pixel by 1. In order to avoid errors, instead of a single digit you can use three: 111 for white and 000 for black. That way, it is much clearer if an error occurs and a single digit has been switched by accident. For example, 010 in a message is most likely meant to be 000. There is no guarantee, of course, since there may have been more than one error in the three-digit string, but this is the most likely interpretation.

To start to see where geometry and symmetry might play a part in building good codes, we can picture the code words 000 and 111 as the coordinates of two opposite vertices of a cube (Figure 76). Any error shifts the code word to another vertex of the cube. To correct an error, we just move the point back to the nearest admissible code word, at the nearest vertex. For example, if we receive 110, the nearest vertex of the cube whose coordinates are an acceptable code word is 111.

But this is a rather inefficient code. First, the code requires us to send three times the amount of data that we are trying to communicate.

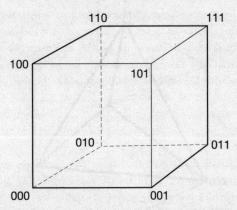

Fig. 76 Code words can be interpreted as the coordinates of the vertices of the cube.

This takes time and energy. When spacecraft were first being launched into space, the machines on board were bulky and batteries weren't powerful, so finding efficient ways to store and transmit data translated into huge financial savings. The first spacecraft to capture pictures of the surface of Mars was Mariner 4, which sent grainy black and white images back to Earth in 1965. Each picture consisted of 4,000 pixels, each of which could take one of 64 different shades of grey. The onboard energy supply allowed only eight pieces of data a second to be transmitted, so it took nearly an hour to send back a single picture. An error correcting code which requires three times the transmission rate is not ideal.

But there is a slightly more efficient code hiding inside the cube. If we embed a tetrahedron in the cube, the corners of the tetrahedron pick out four code words: 000, 011, 110 and 101 (Figure 77). These code words are all the triplets of the cube that have an odd number of 0's. Now, if an error occurs in one of the digits we can detect it because the transmitted triplet will have an even number of 0's. One error causes the code word to slip off the tetrahedron. We don't need the picture of the tetrahedron to understand the mechanism of the code, but it helps to illustrate why efficient codes might be associated with symmetrical shapes. This code associated with the tetrahedron is much more efficient because half the possible messages are code words, yet we can still detect an error. The only downside is that the code can't correct the mistake. The message 010 has an error because it does not

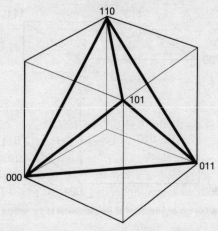

Fig. 77 The code words corresponding to the vertices of the tetrahedron can detect errors.

sit on the tetrahedron, but there are three vertices it could have come from.

The designers of spacecraft communication systems don't want to waste valuable time and energy resending an image of a distant planet because of interference during transmission. Mathematicians realized there were more sophisticated structures that they could exploit to give them a way to actually identify and correct a mistake in the code without having to resend a message.

From error detecting to error correcting

The first clever uses of mathematics to correct codes grew out of the frustrations of an employee at the Bell Telephone Laboratories in America in 1947. Richard Hamming had become increasingly annoyed with the computers at the lab. Several weekends in a row he'd left a program running on the computer only to find on Monday that the computer had dumped his work after detecting bugs in the program. 'So I said damn it, if the machine can detect an error, why can't it locate the position of the error and correct it?'

To see how he went about it, look at this little table:

0	1	1
1	1	0
1	0	

The message consists of four 0's or 1's. These are the numbers in bold. Another four digits at the ends of the rows and columns help to identify any errors – these are called check digits. The power of this code is that it not only checks for errors but also identifies where they are. The digits at the end of a row signify parity – whether there are an odd number or an even number of 1's in that row: 1 for odd, 0 for even, and the same for the columns. If an error occurs in the main body of the message, then the number of 1's changes parity in one row and one column. By looking at the check digits, you can identify where the inconsistency is and the error can be corrected. If an error occurs in one of the check digits rather than the main body of the message, this can also be identified. Two check digits will be wrong if an error occurs in the main text. So if only one of the check digits shows an inconsistency it's that check digit that needs to be corrected.

For example, where is the one error in the following message?

1	0	1
1	0	0
1	0	

The check digits indicate that there should be an even number of 1's in the second row of the main message, but we can see that there are an odd number. So the error is in the second row. To find out which entry is wrong, now look at the columns. The check digits say there should be an odd number of 1's in the first column of the main message, but there are an even number. So the error is the 1 in the second row, first column. The code can now correct this to a 0.

To send a message such as 0111, the four numbers would be arranged in a 2 × 2 grid, the check digits would be added and then the extended grid would be retranslated back into a string of eight digits, in this case 01111010. By adding four extra digits to a message of length 4 and exploiting the mathematics of parity, we can then correct one error that might occur during transmission. Larger messages could be broken down into sequences of four digits, and check digits added to each batch. In this way the code would effectively double the length of the message in exchange for correcting one error per batch of four message digits. But extra digits cost time and money, so the search was on for even cleverer mathematical codes to increase efficiency.

It was the link between codes and geometry that provided Hamming with the key to the discovery of his more efficient codes. A string of three 0's and 1's can be interpreted as a vertex of a three-dimensional cube. A code then consists of a choice of vertices that will be admissible code words. Different codes pick out different shapes inside this cube. For example, the code based on the tetrahedron sitting inside the cube was able to detect but not correct errors. The new error-correcting codes with their check digits are identifying shapes hidden inside higher-dimensional cubes. For example, the error-correcting code described above, with eight 0's and 1's in a 3×3 grid, defines a shape inside an eight-dimensional cube.

Using this geometric insight, Hamming discovered an even smarter way to reduce the number of check digits. For example, one of Hamming's new codes cut the check digits in the previous example from four to three, so that in a string of seven 0's and 1's, four digits contain the message and the other three can be used to correct any error that might occur during transmission. He also extended this principle to apply to longer messages.

At their heart, these codes exploit the geometries that Galois discovered in his investigation of symmetry. Galois had exploited these geometries to produce one of the new families of simple groups of symmetries. Hamming, in contrast, used Galois's geometries to pick out configurations of special vertices in higher-dimensional cubes, and used them as the admissible code words in his new code. Hamming found that this gave rise to lots of code words, but even if a mistake occurred and a code word slipped off one of the special vertices, it was still possible to identify where it came from.

Hamming was keen to get the new codes into print. However, because he was working for a commercial company and the codes clearly had very significant commercial implications, Bell Labs were not so eager to go public. They wouldn't let Hamming release the details until they had got the codes patented. But Hamming was rather sceptical about whether it was possible to patent pure maths: 'I didn't believe that you could patent a bunch of mathematical formulas. I said they couldn't. They said, "Watch us." They were right. Things that you shouldn't be able to patent – it's outrageous – you can patent.'

As soon as he could, Hamming then gathered together all the scraps of paper he'd scribbled his ideas on, and sent them off to the patent office. But in order to register the patent, the office needed diagrams of the switching circuitry to show how error correction would work in practice. Hamming was very much a pure mathematician at heart, and this was well outside his area of expertise. During his time at the lab he had made friends with an engineer called Bernard Holbrook. He used to go down to Holbrook's office and sit and complain about how he didn't fit into a commercial environment. Now, Holbrook immediately saw the power of Hamming's codes and drew diagrams of the circuits that would be needed, signing each page 'witnessed and understood'. Holbrook didn't stop there. Ever the practical engineer, he wanted to see the codes actually working in a real model.

'Since I had the circuit,' recalls Holbrook, 'I turned it over to a technical assistant and said, "Let's build something so we can demonstrate this."' The first reason was to make the patent application even more convincing. 'The second was that I wanted to see how the damn thing would work.' The patent application was successful and was registered on 15 May 1951. It was actually made freely available five years later as part of the settlement in an antitrust case against Bell.

The patent, however, led to a delay in broadcasting the discovery that Hamming had made back in 1948. Because of commercial considerations, he had been unable to tell people outside the company, but he had discussed the new codes with one of his colleagues, Claude Shannon. Shannon was one of the first scientists to suggest a systematic use of 0's and 1's to encode data, and can be called the father of the digital age. Before Shannon, in the so-called analogue age, engineers were wedded to the idea of electromagnetic waves as the way to communicate data. Shannon saw the potential of replacing waves by numbers.

By all accounts, Shannon was a quirky character. While other mathematicians would break for lunch and play mathematical games on the blackboards, he would stay hidden in his office. He only emerged at night, when he could be seen riding up and down the corridors of the lab on his unicycle. Shannon loved inventing things, including a motorized pogo stick, and a unicycle with an off-centre hub which made him bob up and down like a duck when he rode it. He even built a two-seater unicycle, but could get no volunteers to try the thing out. It was Shannon's seminal paper on the power of 0's and 1's that ushered in the modern information age. Deep in the heart of this paper, Shannon let slip a description of the simplest of Hamming's codes, the one that used three check digits to correct errors in messages of length 4.

The delay caused by the patent application and the leak in Shannon's paper allowed another mathematician to slip in and publish his discovery of these codes before Hamming could get there himself. Born in 1902, Marcel Golay was educated in Switzerland before moving to the United States in 1924. His mathematical interests had been stimulated by a club in his hometown that he joined at the age of 15. 'Most of the others were bigger boys and some of them were mathematicians. I remember one in particular of whom it was fun to ask questions. They took pleasure in tutoring a little bit.' But he never had a formal training in pure mathematics.

His other passion was fast cars. He owned several Mercedes, but was deeply frustrated by the sedate driving expected in most American towns. One colleague used to share a ride to work with Golay and described him pulling out of a long queue of cars, driving straight into the oncoming lane, roaring the 100 yards to the lights and downshifting dramatically before taking the turn to the office. After too many close shaves with oncoming cars, his colleague decided to walk to work instead.

It was while Golay was working in the Signal Corps Engineering Laboratories at Fort Monmouth, New Jersey, that he read the account of the simplest of Hamming's codes in Shannon's paper. He quickly saw that this single code could be generalized to a whole family of codes: 'I had been thinking about information theory for quite a while when I was involved with radar. But when I read the paper by Shannon, it was the key because I was ripe for these progresses.' Golay published his discoveries in a paper in 1949 which described these new error correcting codes.

There ensued a bitter dispute about who had discovered these codes first. Although Hamming was the first to have identified them via the geometries discovered by Galois, Golay's paper did include several codes that Hamming hadn't found. One in particular was to prove the catalyst for the discovery of some of the strangest objects to be uncovered on the mathematicians' trek through the jungle of symmetry. It was a set of code words that could be described by strings of 24 0's and 1's. It was an extremely powerful code with many admissible code words. Usually such an increase in admissible code words severely limits the possibility of any error correction. But despite the high number of code words it was possible to correct up to three errors that might occur during transmission and even to detect (but not correct) another four errors on top of these. By interpreting the code words as vertices of a cube in 24-dimensional space, Golay had discovered an extraordinary arrangement of vertices with extremely efficient error correcting abilities. The strange thing was that the code seemed to exist only when you moved into the 24th dimension.

The code proved so efficient that it was used to transmit the beautiful pictures of Jupiter and Saturn taken from the two Voyager spacecrafts in 1979 and 1980. But this code did not only allow scientists to penetrate deep into our solar system. By analysing the symmetries of the code words, mathematicians discovered some of the strangest objects hidden deep in the outer reaches of the world of symmetry. The code words pick out special vertices on a 24-dimensional cube. The group of rotations of the 24-dimensional cube which sends code words to other code words would turn out to be as stunning as any of the pictures from outer space that this code helped transmit.

16 May, Jerusalem

It was a good idea to bring Anton along to the conference. He's been asking lots of questions during the talks. I've seen him in the lobby of the hotel deep in conversation with another conference participant and scribbling in the yellow pads that I got addicted to when I first came to Israel. Listening to the talks, he has been able to place his own work in the context of all the things we mathematicians are so passionate about, which has reassured him of the worth of dedicating himself to the cause. At least for the moment. It has been a really wonderful meeting, a gathering together of my mathematical friends and family. Several of my other former PhD students are here too. Mark has come all the way from South Africa. Pirita, my Finnish student, is continuing her research here in Jerusalem after finishing with me last year.

Also present are all the friends I made here on my first visit in the early nineties, as a 26-year-old post-doc. At that time I too was feeling rather disillusioned with mathematics. I was very worried that I couldn't live up to all the expectations that I felt were being heaped on me after the publication of my PhD. Everyone, I thought, was looking to see what I was going to do next. After the anonymity of being a PhD student, with nobody knowing who you were, suddenly all eyes seemed to be focused on me. But I couldn't prove anything that seemed to match my first epic achievement. Had I burned out already? Was I a one-idea wonder?

It was in Israel that a mathematical colleague explained to me that creativity and discovery run in cycles. Sometimes everything is going well and ideas are flowing. The art of being a successful mathematician is also learning to cope when ideas are not yielding results, when everything leads to a dead end. Once you see that it is a cycle, and things go up again after going down, you can cope with the next down. It was great advice. My time in Israel eventually produced my next up. It was also during that year in Jerusalem that I met Shani. I came back from that visit with a theorem and a wife. Some people make jokes about which will last longer.

Security at the airport as I leave Israel this time is, as ever, high. The two security guards want to know why I've been in Israel. They never seem to believe me when I say 'for mathematics'. 'But you don't look like a mathematician?' I think that means that I haven't got a beard and glasses and look like I've stepped out of the nineteenth century. 'Prove something for us. What's your theorem?'

Many people who go through El Al security hate the invasion of their privacy and resent all the questions. But I love it. To have two people who want to hear about one of my theorems is a rare thing, and I dive into an explanation of how I think you could use Galois's groups $PSL(2, p)$ built from permuting lines, mixed with zeta functions to try to prove that there are infinitely many Mersenne primes, one of the big open problems in number theory. I feel a bit like Sophus Lie when he got picked up by French police on his abortive trek from Paris to Italy. Lie ended up in prison, but I fare rather better with my mathematics and am eventually allowed to board the plane.

I've got quite used to people delving into our lives. When our attempts at IVF failed, we decided instead to expand our family by adoption. That was a more guaranteed route to our goal than the lottery of biology. But being at the mercy of lab assistants with Petri dishes was now replaced by social workers probing into every nook and cranny of our lives. Little did they realize what they were letting themselves in for when they entered our house. For me, it was another chance to explain what it is I spend my life doing to an attentive audience.

We were finally approved as fit to adopt three years ago this May. At the last meeting with our social worker, she asked us whether we would be prepared to adopt siblings. If it meant avoiding going through the long haul all over again, it sounded like a good idea. The waiting game for a child then started.

June: Sporadic

You boil it in sawdust, you salt it in glue
You condense it with locusts and tape
Still keeping one principal object in view –
To preserve its symmetrical shape.

LEWIS CARROLL, *The Hunting of the Snark*

At the beginning of 1900, the classification of the simple groups, the building blocks of symmetry, was shaping up rather nicely. The list of known simple groups went like this: the group of rotations of a prime-sided polygon for different primes p; the alternating groups of degree n, defined by even shuffles of a pack of n cards; and the simple groups of Lie type – groups with a more geometric flavour that Sophus Lie and others had developed. By the 1950s, these groups of Lie type had accounted for 13 different families, all fitting into a nice pattern of groups.

The only slight fly in the ointment of the emerging periodic table of simple groups were five groups of symmetries that had been discovered by the nineteenth-century French mathematician Émile Mathieu. Even set beside all the star names that had passed through the École Polytechnique, Mathieu was regarded as an outstanding student. His tutors were amazed at the speed of his learning. Within eighteen months of arriving at the École Polytechnique, he had already finished the mathematics course and embarked on his doctorate.

It was in 1860, while he was investigating the symmetries in a rather special geometry for his thesis, that Mathieu discovered five new groups of symmetries that were indivisible. He had not been looking for new groups of symmetries, but the mathematics he was studying put him in just the right place and the right time to unearth these gems. Given that the other simple groups discovered by Galois had proved to be the tip of an infinite family of such groups, Mathieu must have thought that this was the beginning of another infinite family. But the more he explored these five little islands, the more it looked as though nothing lay beyond them. No one at the time could make much sense of them. They appeared to be isolated examples in certain geometries that produced indivisible groups. They didn't seem to fit into any pattern of groups, such as those that were emerging from Lie's development of Galois's geometries.

It is curious that Mathieu's unbelievable talents as a young mathematician did not blossom into a spectacular mathematical career. He didn't realize quite how special the groups he had discovered would turn out to be. He drifted into mathematical physics, never to return to the five strange islands he had discovered. Maybe the fact that this little archipelago was not the tip of a whole new continent of groups was disconcerting for someone who loved looking for patterns. Mathieu was regarded by colleagues as rather a shy, retiring figure, and some think that this contributed to his lack of academic success. It might also be that he came upon these groups almost a century before mathematicians were able to give them a context that would unmask their special character.

As the twentieth century unfolded, the feeling grew that mathematicians might have unearthed the whole range of building blocks of symmetry. Leonard Dickson, an American mathematician who had written one of the seminal texts on the 13 families of Lie groups, declared in the 1920s that Galois's group theory was dead. There was a belief that we'd seen everything that was going to be interesting, and the outstanding problems were simply beyond our reach. Apart from Mathieu's five isolated islands, the 13 geometric families, the even shuffles and the prime-sided figures might well be all there was.

This view was further fuelled by William Burnside's belief that if a group has an odd number of symmetries, it can be divided into simple prime-sided building blocks and so cannot produce any new simple groups. If true, this would represent a huge step in determining whether there were any building blocks missing from the list. All the simple groups in the list, discovered by Galois, Mathieu, Lie and others, had an even number of symmetries. If Burnside was right, it meant that any other simple groups would also have to have an even number. But proving Burnside's idea seemed a distant dream. Mathematical problems that are simple to state can sometimes be impossibly difficult to solve.

Burnside's Odd Order Conjecture, as it became known, shifted the focus towards looking for new groups with an even number of symmetries. Burnside believed that Mathieu's groups held the key to whether there were any other indivisible groups out there waiting to be found. After all, there were five anomalous groups, so why not more? Burnside coined the term 'sporadic groups' to describe the rather strange nature of these groups. Despite his suggestions, things went very quiet for a few decades. But in 1954, at the International Congress of Mathematicians in Amsterdam, a clarion call went out to sort out once and for all whether any other simple groups of symmetries were out there.

Lighting the fuse

Richard Brauer had been a professor of mathematics at the University of Königsberg, but because of his Jewish roots his position was terminated under anti-Semitic legislation introduced by Hitler in 1933. Various countries tried to find positions for Jewish academics who were displaced by the new policy, and in 1934 Brauer found himself working in America at the University of Kentucky. It wasn't until he was in his fifties that he made the breakthroughs on symmetry groups that many regard as having launched the final assault on understanding the building blocks of symmetry.

At the International Congress of Mathematicians in 1954, Brauer declared that:

> The theory of groups of finite order has been rather in a state of stagnation in recent years. This has certainly not been due to a lack of unsolved problems. As in the theory of numbers, it is easier to ask questions in the theory of groups than to answer them. If I present here some investigations on groups of finite order, it is with the hope of raising new interest in the field.

He went on to sketch his ideas for how one could limit the possibilities that were available for building a simple group with an even number of symmetries. Combined with Burnside's conjecture that there were no new simple groups of odd order, the outlines of an attack on completing the classification now began to take shape. But many were still sceptical, for the problems looked too difficult for anyone to get to grips with. No one believed that you would be able to confirm Burnside's hunch that the only indivisible groups with an odd number of symmetries were the simple prime-sided shapes.

Brauer, however, had a secret weapon up his sleeve. He had been nurturing the talents of a young graduate student who would turn out to have the determination and the techniques to attack Burnside's Odd Order Conjecture. Born in Austria, Walter Feit had been forced to flee Europe a few years after Brauer. In 1939 his parents managed to get him on the last *Kindertransport*, a train carrying Jewish children out of Austria, just before the Nazis' final round-up of Jews. His parents perished in the Holocaust.

Feit had an aunt who had escaped a few years earlier and found work as a maid in London. Auntie Frieda agreed to look after him, but the severe bombing during the height of the war forced the evacuation of children out of the city. After moving around from one family to another, Feit eventually found himself in a refugee home for boys in Oxford. It was during his time in the university city that the teenage boy became passionately interested in mathematics. When the war ended, his aunt packed him off on a boat across the Atlantic to another branch of the family, based in New York.

In a letter to his aunt, Feit recalled the excitement of the boat trip – not least because there was no rationing on board. 'I had chicken several times,' he reported with glee. There was plenty of food for the hungry teenager, partly because most of the other passengers on the stormy crossing spent their time throwing up instead of eating. When the boat docked in New York on a very foggy winter morning, Feit could hardly see the Statue of Liberty. He was relieved, however, to see his Aunt Regina waiting at the bottom of the gangway.

His new family in America soon took their refugee cousin under their wing and kitted him out for his new life in the States:

I now possess five new pairs of trousers, two new jackets plus new shoes
and lots of new underwear. I have also a watch in my possession . . . I
have been making inquiries about the educational system here and have
been told that my maths is already past college entrance standard.

Feit entered the University of Chicago to study mathematics in 1947.
It was here that he discovered Burnside's great book on group theory
and fell in love with the ideas it contained. He described it as a
'cornucopia' of interesting results. But it was the elegance of both the
statement and proof of Burnside's Two Primes Theorem that Feit
found most attractive: just from knowing that the number of symmetries is divisible by only two primes, Burnside could deduce that
the group of symmetries could be broken down into simple prime-
sided shapes. Burnside became Feit's hero. On the wall of his office he
hung a portrait of the great English mathematician, who thus watched
over his progress. Following Brauer's clarion call, Feit set his sights on
cracking Burnside's Odd Order Conjecture, which would massively
extend the Two Primes Theorem.

Meeting with a like-minded mathematician called John Thompson
was for Feit the beginning of a long, treacherous journey to prove
Burnside's Odd Order Conjecture. Thompson had fallen in love with
group theory after a trip he made to Paris during his studies. The
romantic story of Galois and the ideas he had discovered worked their
magic on the young student as he walked the streets of the city,
retracing the young Frenchman's footsteps. On his return to the
United States, he decided to dedicate himself to the cause of group
theory.

In 1959, while he was working on his doctorate at the University of
Chicago, Thompson spectacularly solved a 60-year-old problem in
group theory. It was immediately clear what an extraordinary talent
had entered the arena. The following year, the university decided to
host a year-long workshop on group theory, to which Feit was invited.
Bringing Thompson and Feit together was like adding potassium to
water: the explosion of ideas that erupted from the two stimulated a
whole new perspective on the theory of symmetry.

The two young students soon discovered that Burnside was a
common hero. Together, they gradually began to work their way
towards proving Burnside's Odd Order Conjecture. They estimated

that it might take about 25 journal pages to set out the ideas they were hatching for a proof. But as they got sucked further and further into the problem, they saw that their initial estimate had been rather optimistic. Thompson describes how, even now, the problem looks as tough as granite. And yet they somehow had the confidence to keep going even though the proof they were constructing was growing ever longer and more complex. 'It was a tough problem and probably imprudent to even try it, but how else should one behave in one's youth?' Thompson describes how their thought processes became inextricably intertwined during that year.

The other group theorists who were part of the 1959 workshop in Chicago would enjoy congregating at tea to see how much progress had been made. Many believed that it was just too difficult. Even Brauer, who initiated the attack, was not really convinced that just knowing there were an odd number of symmetries could possibly be enough to pull the group of symmetries apart: 'Nobody had any idea how to get started. It was not even clear that the whole problem made sense.' What did being divisible by 2 or not have to do with the internal structure of the symmetries?

Gradually, Feit and Thompson ground the problem down, solving various stages on the way to their main goal. The smaller successes gradually built up their mathematical muscles, making them feel strong enough to mount the final assault. Finally, they believed that they had all the details in place. They were ready to announce their epic proof of Burnside's Odd Order Conjecture:

> It was technical – there was no way to avoid it. But it was a wonderful thing. We'd finally busted it. But then, just before we were about to submit the paper, Walter noticed a mistake. If Walter had not found the gap, I almost certainly would not have found it; we would have submitted a flawed manuscript and eventually someone would have blown the whistle. If that had happened, it is doubtful that we could have generated a new head of steam to bust the difficulty, which in fact took us several additional months of thought and nail biting.

Eventually, they fixed the problem. When they finished writing up the complete proof, they had a paper which ran to over 250 pages. Mathematics had never before seen a proof of such complexity and length. The first few journals they sent it to didn't know what to do with it. Papers were usually 30 pages long. It was rejected several times before the *Pacific Journal of Mathematics* had the courage to dedicate a whole issue to publishing the proof.

The publication, in 1963, changed the face of the subject. Some have described the occasion as comparable to the moment in evolution when fish emerged from the water onto dry land. It inspired a whole generation of young mathematicians who were eager to follow Thompson and Feit out of the swamp and into the exciting new world their proof had opened up. The gauntlet was thrown down to see whether we could finally understand the full range of the indivisible groups of symmetries that Galois's work had begun. But students would find that it took them a whole year of their research just to work their way through the 255-page proof in the *Pacific Journal*. Thompson is still sceptical whether, even four decades later, many have really appreciated the precision and subtlety that went into weaving together a proof of such intricacy.

Many of the young researchers who set out to battle their way through the paper believed that it contained ideas that could be used to complete the periodic table of symmetry. Many believed that it was now just a matter of showing that the prime-sided shapes, the alternating shuffle groups, the 13 families of Lie and the five strange sporadics of Mathieu were all you needed to build it. But around the same time as Feit and Thompson cracked Burnside's Odd Order Conjecture there came a warning shot across the bows: maybe there was more to the periodic table than had been thought.

A Japanese mathematician, Michio Suzuki, discovered a completely new infinite family of indivisible groups. Then a Korean mathematician, Rimhak Ree, followed up with two other infinite families. The revelation of these new families was initially deeply shocking for mathematicians who thought they knew the lay of the land. But it soon became clear that these three new families were special versions of Lie groups and could therefore be safely taken under the familiar umbrella. All that had happened was that the Lie groups had expanded from 13 families to 16. Nevertheless, psychologically, it rocked mathematicians' faith in their roadmap of how things were going to pan out.

A few years later, it transpired that they had had every reason to doubt themselves. Thompson received an unsettling letter from a Croatian mathematician working in Australia. The revelation contained in this letter could not so easily be assimilated into the status quo.

Janko's first bookend

Zvonimir Janko had used his mathematics to escape the increasingly repressive regime of his native Yugoslavia, and had landed up in Canberra after a detour via Germany. Like many others of his generation, he had been enlisted to the cause of group theory by the epic opus of Feit and Thompson. Thanks to their proof of the Odd Order Theorem, any new indivisible simple groups out there had to consist of an even number of symmetries. Janko began to explore the implications of what this would mean for the structure of the group of symmetries. The expectation was that you would always be able to show that the group had to be either a group of shuffles or the symmetries of one of Lie's geometries. But Janko had come up with a strange case that didn't fit into Lie's framework, and he couldn't see how to explain this example away.

Thompson had developed a formula which suggested possible sizes for indivisible groups with an even number of symmetries. Using it, Janko had discovered that there might be an indivisible group with 175,560 different symmetries. What worried him was that none of the existing indivisible groups that had been discovered had this particular number of symmetries. At this point, Janko's usual strategy was to find an argument to show why this number of symmetries was impossible. But all the tricks that had knocked out other possibilities weren't working. Increasingly, he began to suspect there really was a simple group with this number of symmetries. And if it existed, it was going to be a completely new object.

Janko knew that Thompson would appreciate what he thought he'd discovered. Thompson's first impression on receiving the letter from this unknown mathematician was that there must be a simple way to show the simple group didn't exist. After firing off a letter to Australia in which he said why he thought it couldn't exist, Thompson realized he'd made a mistake. When he came into the department in Chicago the next morning he was wearing a rather serious look. He couldn't make this group go away – but did it really exist? If it did, Janko knew how many symmetries it would have, but it was another matter entirely to construct a geometric object with an indivisible group of 175,560 different symmetries.

Janko and Thompson started corresponding. Thompson still thought that the group didn't exist, but, he remembers, 'Janko just stuck to his guns.' Janko came over to the States and started giving talks about the possible existence of this new group. Many were sceptical. In talks they would question whether he had used all the usual techniques which generally proved why such a group couldn't exist. Janko was utterly confident that there was something new out there. He would describe how beautiful his hypothetical group of symmetries would be, if only he could construct a geometric setting in which to realize it. If his mathematical powers of persuasion failed, he would silence his doubters with the simple statement 'My group is safe . . . because Walter Feit has stopped trying to find contradictions.'

Janko finally made the breakthrough that silenced his doubters for ever and ushered in a new era in the theory of symmetry. The mathematics had acted like a radar, telling him where this new island should be. In 1965 he finally discovered the geometric setting he was looking for: one whose group of symmetries was precisely what he had predicted. Galois had discovered the first simple Lie group by constructing a geometry of two-dimensional lines based on the finite number system $0, 1, \ldots, 6$. Using a shape in seven-dimensional space built from a number system with 11 elements, $0, 1, 2, \ldots, 10$, Janko finally had a new simple group with 175,560 symmetries that couldn't be divided into smaller symmetries. 'I am now the world's expert on seven-dimensional space over the field with 11 elements,' he told Thompson. Janko was deservedly proud of his discovery and had no qualms about calling the group J.

He published his discovery in 1966. In contrast to the 255 pages Thompson and Feit had needed for their proof, Janko's paper took up a single page, but the implications were potentially as far-reaching. A hundred years earlier, Mathieu had constructed his five anomalous groups, groups that Burnside had christened sporadic. Mathematicians had quietly stuck their heads in the sand, hoping that these groups wouldn't spoil the beautiful classification of the building blocks that Galois and Lie had laid out for them. But suddenly, here was a sixth strange building block which popped out of nowhere. The implications were clear: if there could be a sixth, why not a seventh, an eighth . . . where would it end?

Indeed, Janko had got a taste for discovering these new symmetries, and used his strategy to predict two more indivisible groups. The first group, J, became J_1. Now he had J_2, with 604,800 symmetries, and J_3, with 50,232,960 symmetries. Again, the numbers were there before the construction of the actual groups. But this time it wasn't Janko who constructed them. The second was constructed explicitly by Marshall Hall, Jr, who found a group of special shuffles of 100 cards that was indivisible and had exactly 604,800 shuffles in it. The third Janko group was constructed by Graham Higman and John McKay in Oxford.

There began a rather tense stand-off that got the whole group theory community talking. Who should get the credit for the group – the person who predicted the existence of a group with this many symmetries, or the person who went out and found it? With his first discovery, Janko had done both these things and so there was no quibble about the group being called J. But with these two new groups Janko was predicting, things were a little messier. Hall would get rather upset if the group he'd constructed was called simply the second Janko group. What about credit for the work he'd done in actually making this group a reality?

Janko soon found that he wasn't the only one in the game of predicting these new sporadic groups. More and more young guns saw the potential for making their name and getting a group named after them. Discoveries came fast and furious. Donald Higman and Charles Sims constructed their group over dinner – inspiration struck between the main course and dessert. They were visiting Oxford as part of a conference on group theory, and the conference dinner was being held in one of the colleges. The staff were clearing the decks for the next part of the meal while guests took a stroll around the quad. By the time they completed the fourth edge of the square, Higman and Sims knew that they had the tricks to construct another new indivisible simple group with 44,352,000 symmetries. It was a different group of shuffles of 100 cards which, like Janko's second group, turned out to be indivisible.

Jack McLaughlin found another indivisible group with 898,128,000 symmetries. Dieter Held, a colleague of Janko's in Australia, identified a possible simple group with over four billion symmetries which was then constructed on the other side of the planet by Graham Higman and John McKay in Oxford. By 1968, Michio Suzuki, who'd shocked the world with his new infinite family of Lie groups, had constructed an 11th sporadic group with nearly 500 billion symmetries.

An important tool in the actual realization of these groups was the computer. Many of the calculations that had to be performed to confirm the existence of these groups were becoming too protracted to be solved by other means. Like a telescope, the computer was allowing mathematicians to stare farther and farther into the deep space of symmetry, and every now and again another isolated planet was picked up in its gaze. But it still required the mathematician's strong intuition about the coordinates at which to point the telescope.

With so many new sporadic groups appearing, the whole thing was starting to look like a nightmare ragbag of exceptions. The question of where and whether it would ever end started to creep into many people's minds. A song written at the time captured the growing sense of panic that was beginning to grip group theorists:

> The floodgates were opened! New groups were the rage!
> (And twelve or more sprouted, to greet the new age.)
> By Janko and Conway and Fischer and Held
> McLaughlin, Suzuki, and Higman, and Sims.
>
> No doubt you noted the last lines don't rhyme.
> Well, that is, quite simply, a sign of the time.
> There's chaos, not order, among simple groups;
> And maybe we'd better go back to the loops.

It was then that a new group entered the game which gave a sense of unity to many of the groups that had sprouted up since Janko's first group in 1965. The discovery of this group brought one of the most colourful players into the story of symmetry.

'When I grow up . . . I want to be a mathematician'

It has often been asked whether great mathematicians are born or whether they are nurtured. On the whole, nurture counts for more than nature, but every now and again someone comes along whose innate mathematical ability seems to have been there from birth. It's as though John Conway's brain is hard-wired for mathematics.

Conway was born on Boxing Day 1937, and it wasn't long before his mathematical skills started to become evident. His mother discovered him at the age of four reciting increasingly higher powers of 2 to himself. He excelled at everything at primary school. When he was asked at the age of 11 what he wanted to be when he grew up, he already had his destiny clearly fixed in his mind: 'I want to be a mathematician in Cambridge.'

Conway attended the local grammar school in Liverpool, where his father, Cyril, was a chemistry laboratory assistant. During the war, when other teachers were called up, Cyril Conway got the chance to teach chemistry full time. (Later, he would teach two members of the Beatles their chemistry.) He was keen to communicate his enthusiasm for science to his son. On one occasion, in order to impress his son with the magic of radio waves Cyril rigged up a radio so that it had strands of silk coming out of the back. Then, as the radio played away, he ceremoniously cut the silk to show the boy that it didn't need them. John was amazed that he could still hear the music. He was also fascinated by a telephone network that his father set up between the air-raid shelters during the war, and constructed his own telephone network that he used to communicate with his friends.

Conway decided he would master the secrets of numbers in the six months before he was due to head off to start his degree at Cambridge. He set himself the task of being able to factorize into their prime building blocks any number below 1,000. Arriving in Cambridge was something of a shock for him. 'I found it very hard because most of the students were from rather posh homes and I was a poor boy.' Eventually he found a set of friends who, regardless of their social background, appreciated the speed with which he could factorize 999 into $3 \times 3 \times 3 \times 37$.

During one summer vacation, when he was working to earn money in a biscuit factory, Conway added another feat of memory to his repertoire. His job consisted of cleaning the soot off the ceiling of the huge oven in which the biscuits were baked. He would scrub for hours, but the ceiling remained black. Despite the futile nature of the job, Conway needed the money, so to relieve the boredom he decided to learn the decimal expansion of π. In the late 1950s, π was known to 808 decimal places. By the end of his holiday, Conway could recite all 808 places. As more decimal places have been calculated, Conway has extended his range. Of course, he was fully aware that learning the decimal expansion of π was mathematically as futile as trying to clean the soot off the biscuit factory oven. He soon turned to more fundamental questions about numbers.

By the early 1960s Conway was on his way to completing his doctorate and realizing his dream of becoming a mathematician at Cambridge. He was supervised by a number theorist, Harold Davenport, who had set him an extremely tough problem: to prove that every integer can be written as a sum of 37 numbers, all of which are fifth powers. This conjecture had been made nearly two hundred years earlier by another Cambridge mathematician, Edward Waring. Conway and Davenport held a regular meeting every Wednesday at 11 a.m., but Conway was rather disorganized and always arrived late. Davenport, the model of a Cambridge gentleman, would always say, 'No problem, I only arrived myself just a few minutes ago.' When Conway missed an appointment completely, he sent his supervisor an apologetic note only to receive a reply by return: 'No problem. It was such a nice day I took my wife and children to the seaside and forgot myself.'

One Wednesday, Conway turned up declaring that he'd cracked Waring's problem. Davenport was incredulous, but after checking the solution he couldn't find any mistakes. 'Mr Conway, what we have here is a poor PhD thesis.' Conway was crestfallen. But then he realized Davenport was actually giving him the green light to pursue the problems that really interested him, and he started to move towards logic and set theory.

While the early 1960s was a time when Cambridge students still looked quite smart and civilized, Conway looked like he'd arrived five years too early for the hippie movement that would engulf Cambridge in 1968. He'd been sporting long hair and sandals since the age of 14, and had added a full ginger beard to the ensemble. He was more interested in doing mathematics than his own appearance.

Some years later, on the way to a conference in America at the height of the hippie movement, Conway was stopped at customs by officials who were convinced that the length of his hair made him a prime suspect for smuggling drugs. They pulled him aside and rifled through his bags until they triumphantly found what they were looking for: a tin with the lid taped down and a label declaring the contents to be something called 'Marvel'. They were convinced this was where the long-haired hippie had hidden his drugs. He was asked to open the tin. Conway obliged and tipped the contents onto the desk – five beautifully made Platonic solids tumbled out. He had crafted them from cardboard and sealed his symmetrical creations into the tin to protect them during transit. Eventually Conway was allowed to enter the country after he'd convinced the officials that the dodecahedron and icosahedron were not secret containers for a stash of drugs.

As a graduate student, even his love life revolved around mathematics. Conway met his first wife, another mathematician, while he was researching his doctorate. Mathematics was never too far from their courtship. Rather than whispering sweet nothings to each other, they would go for walks along the river in Cambridge and recite the expansion of π, taking it in turns to do 20 decimal places each.

After three years of research, Conway completed his thesis and the time came for his viva. Usually this oral examination is a very formal affair conducted in a lecture room lined with blackboards. But Conway's thesis was of such exceptional quality that the examiners did not feel the need to cross-examine their witness in such harsh surroundings. Instead, Conway's viva took place in the Fellows' Private Garden. Since none of the examiners had a key to the locked gate, Conway obliged by picking the lock with a paper clip.

Having successfully got his degree and doctorate at Gonville and Caius College, Conway was elected to a fellowship at Sydney Sussex College. Despite realizing his childhood dream, Conway was becoming rather dejected. He felt restless, concerned that he hadn't really done anything of great importance. 'I became very depressed. I felt that I wasn't doing real mathematics; I hadn't published, and I was feeling very guilty because of that.'

He spent a lot of time in the department playing backgammon. But that didn't really help his mood. 'I used to feel guilty in Cambridge that I spent all day playing games, while I was supposed to be doing mathematics.' Suddenly the economy took a downturn, and other more worthy mathematicians were finding it difficult to get jobs. That only made Conway feel even more guilty. He had a cushy job at a Cambridge college while unemployed mathematicians were producing work that he regarded as far superior to his own. But he knew that he deserved his position, and that he was capable of producing world-class mathematics. He would need to justify himself and prove something really important.

Conway got his lucky break in Moscow in 1966. It was the summer of the International Congress of Mathematicians, the four-yearly mathematical jamboree at which the subject's big prizes, the Fields Medals, are handed out. That year one prize would go to an enigmatic mathematician called Alexandre Grothendieck, who refused to come and collect his prize as protest at the increasing military escalation in Russia. Conway, however, was simply enjoying his first taste of a big congress. It was while he was spending the day at the Moscow centre for symmetry, appropriately enough, that he met John McKay. Conway was manning a stall handing out bread rolls stuffed with meat when McKay strolled up to him. In return for the roll, McKay handed Conway the gift that would change his life.

'There's a rather interesting symmetrical object I think you might be interested in,' said McKay, and, with a mouth full of bread roll, he launched into a description of something called the Leech lattice that he thought might have an interesting symmetry group. Just two years before the Moscow congress, an English mathematician called John Leech had discovered a rather spectacular geometric arrangement of 24-dimensional spheres.

24-dimensional grocers

If a grocer stacks tins of soup on a shelf, they are usually lined up in rows and columns so that, from above, the tins are arranged in what is called a square lattice. But if the grocer wants to get as many tins on the shelf as possible, then this is not the most efficient way to do it. If you take a collection of marbles and let them settle in the bottom of a wok, then they will arrange themselves into a perfect hexagonal pattern, the most efficient packing arrangement.

The mathematics that explains why the hexagonal lattice is the most efficient way to arrange circles was discovered by Joseph-Louis Lagrange, who had inspired Ruffini, Abel and Galois. His mathematics proved that if you want the pattern to be regular (repeating itself left and right, up and down), then you can't beat a hexagon. The circles will cover a total of just over 90 per cent of the shelf if you use this lattice. The exact fraction of space occupied is π divided by the square root of 12. In contrast, the square lattice covers only about 78 per cent ($\frac{1}{4}\pi$) of the surface.

Lagrange actually had no idea that this is what he had proved. The result followed as a consequence of work contained in a treatise he'd written about arithmetic and equations, and it was Gauss, with his geometric intuition, who understood that Lagrange's mathematics had this physical interpretation. In a review of Lagrange's opus, Gauss explained why Lagrange's calculation implied that the hexagonal lattice was the most efficient way to cover a two-dimensional surface with circles if the pattern was regular.

But was there a way to better the hexagonal lattice if the pattern was allowed to be irregular? For another hundred years mathematicians tried to see if there might be a way to arrange circles more compactly if they were allowed to be put together in a chaotic fashion. After a number of false starts, it was the Hungarian mathematician László Fejes-Tóth who proved in 1940 that symmetry is best: no irregular arrangement of circles can beat 90 per cent.

Once the grocer has stacked the cans of soup and moves on to stacking oranges, the problem cranks up a dimension. Instead of two-dimensional circles, the question now is about finding the most efficient way to pack three-dimensional spheres. Here the grocer generally goes for what is now known to be the most efficient way to arrange the oranges. You start with a layer of oranges laid out in a hexagonal configuration. Then on top of these you place another hexagonal layer, so that each orange in the second layer nestles between three of the oranges in the first layer. Keep on repeating this as the tower of oranges grows. Each orange touches another 12 oranges: six in its own layer, and three in each of the adjacent layers.

The amount of space occupied by the oranges comes out at about 74 per cent (π divided by the square root of 18). In 1661, Kepler conjectured that this was the best the grocer could do. Gauss proved in 1831 that no regular arrangement of oranges could beat this fraction (something called a lattice packing). Incredibly, it took till 1998 before a mathematician and a computer could prove that no irregular arrangement could somehow beat the 74 per cent conjectured by Kepler.

Thomas Hales of the University of Pittsburgh showed how one could reduce the problem to a finite but large number of calculations that would confirm that the grocer's symmetrical stack of oranges was the most efficient. The calculations that Hales needed to do were implemented and checked on a computer. The fact that a computer was used to prove Kepler's Conjecture unsettled some people in the mathematical community. But most recognize that the brilliance of Hales's proof is in showing that there are a finite number of configurations that need to be checked in order to confirm Kepler's hunch – after all, intuitively one might think that allowing for any irregular arrangement leads to an infinite number of options. But Hales's proof shows why symmetry wins again.

For our grocer on the corner, this is where the problem ends. Mathematics has helped him to stack tins by calculating with two-dimensional circles and to stack oranges by calculating with three-dimensional spheres. But a mathematical grocer can't resist asking, 'What about four-dimensional oranges?'

To investigate a four-dimensional orange we can no longer rely on a geometric picture: instead, we must describe it using numbers. Draw a circle of radius one unit on a piece of paper (Figure 78). Each point on the circle can be identified by two numbers, the coordinates of that point on the map. We can write down an equation which describes all the points (x, y) on the circle: the coordinates of the points must satisfy the equation $x^2 + y^2 = 1$. This equation is therefore another language for describing the geometry of the circle.

Points on the surface of a three-dimensional sphere need three coordinates to identify them. Again, we can translate geometry into an equation. A point (x, y, z) will lie on a sphere of radius 1 if its coordinates satisfy the equation $x^2 + y^2 + z^2 = 1$. Although the pictures run out when we move up to a four-dimensional hypersphere, the

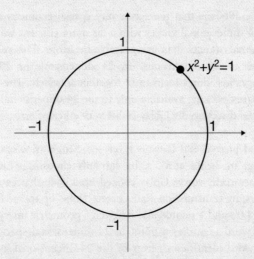

Fig. 78 The circle consists of all those points with coordinates (x, y) satisfying
the equation $x^2 + y^2 = 1$.

language of equations is quite capable of keeping up. So although a
four-dimensional hypersphere can't be drawn or built or grown in
our three-dimensional world, it can be described. Each point on the
four-dimensional hypersphere is described by a set of four numbers
(x, y, z, w) which satisfy the equation $x^2 + y^2 + z^2 + w^2 = 1$. So the equations and numbers help us to 'see' a four-dimensional ball, even though
we will never be able to physically construct it.

Although we can't see these balls, thanks to the language of equations
and numbers we can still arrange them in symmetrical ways and
measure how much space they take up. The mathematical grocer has
discovered that the most efficient way to stack four-dimensional
oranges is to generalize the hexagonal packing in three dimensions.
One description of the hexagonal packing in three dimensions is that
the oranges are positioned such that their centres all have coordinates
(x, y, z) with $x + y + z$ an even number. So the four-dimensional grocer
just generalizes the pattern and arranges the four-dimensional oranges
such that the centres of the oranges are located at points (x, y, z, w)
where $x + y + z + w$ is an even number. It turns out that in four dimensions you can't beat this configuration either.

The exciting revelation that the British mathematician John Leech made in the 1960s is that there is a very special geometry that provides the 24-dimensional grocer with a far more efficient way to pack 24-dimensional oranges than using simple hexagons. This geometry is very special to 24 dimensions. In 23 dimensions or 25 dimensions, oranges just don't seem to fit together so nicely. The existence of this amazing packing available only to the 24-dimensional grocer is related to the discovery by Golay of his very efficient error correcting code.

Golay had proved that there is a very efficient way to encode data using strings of 24 0's and 1's. By carefully choosing which of the 2^{24} strings are code words, Golay cooked up a code that could detect seven errors in transmission and correct three of them. But these strings of 24 0's and 1's could also be given a geometric interpretation. Each code word identifies a point in 24-dimensional space. In fact, each code word identifies a vertex of the 24-dimensional hypercube. Leech discovered that these code words are the secret to how you should arrange the 24-dimensional oranges in the most efficient manner.

We can see the principle at work in three dimensions. Consider the code with four admissible code words: $(0,0,0)$, $(1,1,0)$, $(1,0,1)$ and $(0,1,1)$. These are the strings of three 0's and 1's with the property that there are always an even number of 1's in the word. If we get a message containing a triplet with an odd number of 1's, we know that an error has crept in. So this code can detect but can't correct an error. But as we saw earlier, these code words identify points on a cube. If we put oranges with centres at these points and repeat the pattern, we actually get the hexagonal packing that the grocer uses to stack oranges. So code words tell you the best place to put your oranges.

This is why the discovery of the Golay code was key to building Leech's efficient packing of 24-dimensional oranges. The secret of the efficiency of the hexagonal packing in three dimensions is its symmetry. Leech realized that the symmetry of his extraordinary arrangement in 24 dimensions should also warrant more attention. Because he was not sufficiently well versed in the language of group theory to be able to explore the symmetries, he set out to find someone who might be able to analyse them. He dangled the 24-dimensional oranges in front of various group theorists in Oxford, but no one took the bait. John McKay, like Leech, could sense that the symmetries of the packing had the potential to reveal something interesting. So he joined in the pestering of the group theorists to try to get them to look at this problem.

It was a daunting prospect. Understanding the group of symmetries meant understanding the different ways in which you could move the arrangement of 24-dimensional oranges around so that they would magically all realign. For the hexagonal packing in three dimensions one could easily pick out symmetries – for example, rotating by a sixth of a turn around one of the oranges. This would rotate the stack such that the oranges all magically realigned. But spinning 24-dimensional oranges was going to require an extraordinary feat of mental gymnastics. When McKay saw Conway in Moscow, he wondered whether he might be the mathematical magician who could juggle this symmetrical stack of balls and see what was going on.

The treasure chest symmetry group

Conway was certainly intrigued. When he got back to Cambridge, he started exploring Leech's packing and began to see what all the fuss was about: the geometry was full of symmetry. It was like looking at some extraordinary Moorish design in a 24-dimensional Alhambra and wondering whether this was a genuinely new symmetry or an old symmetry in a new guise. Even so, the task of understanding the symmetries looked monumental.

Cambridge had just succeeded in tempting one of the great group theorists of the day to take up the offer of a professorship. This was John Thompson, who had worked with Walter Feit on that epic paper which, as far as group theory was concerned, changed everything. Conway naturally thought that Thompson would be the ideal person to sort out what the symmetries of the Leech lattice looked like. He kept badgering Thompson to have a look at it, but it became obvious that the new professor wasn't going to take up the challenge. The trouble was that having secured his reputation as the world's greatest group theorist, he was now being approached by anyone who had a crazy idea for a new group of symmetries, expecting him to sort out their problems. Conway admits that 'most of the ideas were just junk. It wasn't worth his while to devote a lot of time sorting these ideas out and find they probably wouldn't work anyway.' But Conway was confident that the Leech lattice was more than just junk.

He kept nagging away at Thompson: 'Every now and again I'd ask him whether he'd thought about it, and he'd never have thought about it. But he gave you the impression it was just because he hadn't got round to it.' Eventually, after bearding Thompson yet again in the common room, Conway asked him outright, 'You're not going to think about it are you?' 'No' was Thompson's reply. Seeing Conway's disappointment, he added, 'Look, you find the size of this group, then I'll be interested. Until then I'm not.' And he walked off.

In contrast to what had happened with the sporadic groups discovered recently, Conway already had the geometric set-up before he knew how many symmetries there were or whether the group of symmetries would even be indivisible. Thompson was used to seeing things in the other order: first someone would use his formula to make a proposal for the number of symmetries in some new candidate sporadic group before anyone went off trying to find a geometry with that number of symmetries.

When Conway returned home that evening, he was on a mission. Life was tough. He had a wife and three children under the age of four to support and was having to do extra teaching to make ends meet:

> I had to work like stink to earn enough money 'cos I didn't have a proper job yet. So I was working like all hell. But I decided the symmetries of this thing were important and I'd better make some time for doing it.

So he formulated a plan of action: he would spend 12 hours every Saturday from 12 noon to 12 midnight working on it; then on Wednesday evenings he'd put in another 6 hours from 6 p.m. to midnight. His wife wasn't too pleased with the idea as it was going to take him away from the children at the weekend. 'I told my wife if I could do it that it would make my name. It was a really big deal.' She could see this was important so reluctantly agreed.

The first Saturday came. At 12 midday on the dot, Conway ceremoniously kissed his wife and kids goodbye as if embarking on an Antarctic expedition, and locked himself in the front room of their house. The house had been a derelict wreck before a developer had bought it and done it up and sold it to the Conways. There was still lots of junk lying around the place from the renovation. In a corner of the front room Conway found a roll of wallpaper, and decided that this was the perfect canvas on which to log his exploration of this strange new symmetrical object. He laid the roll on his lap and started to write down everything he knew about this 24-dimensional geometry.

After about three hours he could see that he'd been making far too many guesses about the structure. One of them had to be wrong, because they were starting to contradict each other. So he decided to start all over again. He allowed himself one guess only: that there was some hidden symmetry which was not obviously identifiable. From this one guess he started to logically piece together how many other symmetries there must be in this geometry. Everything had to be watertight. By 6 p.m. he'd worked out how many symmetries this configuration must have based on the one guess he'd made.

Conway remembered Thompson's promise that he'd start listening once he knew how many symmetries there were. It was a staggering number: 4,157,771,806,543,630,000 symmetries.

> Or double that, I still wasn't completely sure. But I thought it was close enough to pick up the phone to Thompson. I just told him the size, and my God he was interested. About 20 minutes later he phoned me back.

Thompson told him that if Conway was right about the size of the group it would indeed be a new sporadic simple group to add to the list. The symmetries of this configuration would have to be so inextricably linked that, as with the symmetries of the icosahedron, there was no way to build it by putting together smaller symmetries. It was as indivisible as if it were a prime number. But that wasn't all.

In that 20 minutes, Thompson had worked out that this group would have to contain as sub-symmetries nearly all the other sporadic simple groups that had been found. A group can be indivisible but still contain subgroups of symmetries. It's just that when you try to divide the indivisible group by the subgroup the result is not the symmetries of another symmetrical object. For example, although the 60 rotational symmetries of the icosahedron are indivisible, the rotational symmetries of a triangular face sit as a subgroup of symmetries inside this indivisible symmetry.

In a similar way, the huge indivisible group of symmetries of the Leech lattice contained an array of sub-symmetries. Thompson had worked out that it would have to contain the five groups of Mathieu, the second Janko group, the Higman–Sims group, the McLaughlin group, the Suzuki group and another two previously unidentified groups. Discovering the symmetries of the Leech lattice was like opening a huge box and finding it full of treasure. Conway was on the verge of unearthing three new indivisible sporadic groups. Not only that: since the symmetries of the Leech lattice contained all these other sporadic groups, it seemed to provide some overarching logic to explain what had previously looked like a hotchpotch of unrelated groups. However, he wasn't home yet. The size of the group had been calculated by postulating the existence of some hidden symmetry. Conway was convinced that it must be there somewhere, and to claim his groups he would first have to find it.

Conway still had five hours left of his first Saturday trek. After three more strenuous hours he came up with a symmetry that might provide the missing link. He called Thompson again. He still had to check that the effect of the symmetry he was proposing would align all the balls in the geometry as perfectly as he hoped. But, he told Thompson, he was exhausted and he was now going to bed. He'd do the checking on Sunday morning. He felt he was close, and that he could probably steal some more time away from the family.

But once he'd put the phone down, Conway just couldn't resist pushing the thing a bit further. By now the roll of wallpaper was full of his calculations. After a while, he identified 40 additional calculations that he would have to do to confirm the hidden symmetry. He phoned Thompson again, just before 11 o'clock, and said that he hadn't gone to bed after all. He'd narrowed down what he needed to do to the 40 calculations that would test his hunch – but this time, he said, he really was going to bed. He was shattered.

Mathematics takes a very powerful hold of you once you've caught the bug. Conway decided to see how long one of the 40 calculations would take, so he timed himself. When he put his pen down, two minutes had passed. Not bad – it would take only an hour and 20 minutes to check the whole thing. At 12.20 Thompson's phone rang for the fourth time that night. It was not hard to guess who would be phoning him so late. An excited Conway said, 'I've done all 40 calculations. It checks out. I've found my missing symmetry. This time I really am going to bed.' They arranged to meet the next morning to go through Conway's epic journey.

Conway had gone 20 minutes over the time he'd allotted himself for his first Saturday journey. But he wouldn't need the six hours on Wednesday, nor any of the months of other Saturdays and Wednesdays he'd set aside for the project. He'd found his group! He went up to bed, but the excitement of the last 12 hours made it impossible for him to sleep, and he slipped back down to the front room to look once more at the roll of paper.

The discovery of this new group had a powerful effect on Conway's psyche:

> I knew I was a good mathematician but I hadn't done the work to prove it. I'd been feeling really black for several years. I felt really guilty about the amount of time I spent playing backgammon in the department. Serious people would walk past me giving me disapproving looks. The discovery of this group wiped out that guilt. It removed the black feeling.

It also propelled Conway into the mathematical jet set. He was invited all over the world to explain the discovery of his treasure box group that contained so many wonders. 'I even flew over to New York, gave a 20-minute talk then flew back that afternoon.' He had three new groups to his name: Conway 1, 2 and 3, as they became known. In fact, if he'd got straight down to the job after coming back from Moscow, he could have made it seven.

What McKay had done was hand Conway a twentieth-century dodecahedron. When the Romans had shown Pythagoras the crystals of pyrite that inspired his discovery of the sphere of 12 pentagons, the Greek mathematician must have realized that he was holding a mathematical gem, although it took until the nineteenth century and the genius of Galois to reveal the importance of its symmetries. For Conway, McKay's gift would be just as rewarding. Its contours would cement his name in the roll call of symmetry.

Conway spoke about his discovery for the first time in Oxford, at a seminar organized by the department's guru of group theory, Graham Higman. McKay, who was working at the Atlas Computing Laboratory at Chilton, just outside Oxford, regularly attended Higman's seminars. The two of them had successfully constructed some of the growing number of sporadic groups by exploiting McKay's computing abilities. McKay was especially excited to hear whether the seed he had sown in Conway's mind in Moscow had come to fruition. The result was more than he could have hoped for. To see the symmetries of this large lattice of 24-dimensional oranges binding together 12 of the strange sporadic groups of symmetries was mind-blowing.

Conway stayed the night at McKay's house. In the middle of the night, McKay burst into Conway's room in a state of agitated excitement. 'What you've discovered is one of the deepest secrets of the universe!' He launched into a frantic description of how important Conway's discovery was. McKay's wife came in and tried to calm him down, but nothing would bring him down from the high he was on. Eventually, Conway noticed, his wife slipped him a sedative. She explained that mathematics quite often had this effect on her partner.

Fischer's phoenixes

The Leech lattice that McKay had shown Conway that afternoon in Moscow was an extremely efficient packing of balls that worked only in 24 dimensions. Conway's analysis of the group of symmetries of this arrangement was therefore something rather special. Since the packing was unique to 24-dimensional space, this group was not going to fit into a nice new infinite family of groups like the Lie groups and the shuffles of a pack of cards. It was what Burnside had dubbed a sporadic group. And it was huge, with over four billion billion different symmetries. But these gigantic sporadic groups didn't stop there.

A year after his discovery, Conway learnt that a German mathematician had beaten him by finding three more strange sporadic groups that were larger than Conway's – by a factor of 30,000. Bernd Fischer remembers meeting Conway for the first time at a meeting at Oberwolfach, the research centre in the middle of the Black Forest where I'd first met Fritz. Fischer was sitting in the centre discussing mathematics on the afternoon of the first day when he saw a wild-looking man emerge from the woods clutching a huge role of wallpaper. One of the other mathematicians recognized Conway and reassured Fischer:

'Don't worry – he's a scholar.' When Conway was introduced, instead of shaking hands he shook his fist at Fischer: 'You beat my group!'

Fischer almost missed his groups completely. They had their genesis in the library in Frankfurt where he was a student. His supervisor didn't trust librarians – 'They always mess up all the books' – so he'd arranged instead for the students to run the library, taking turns to do one-hour shifts. It was during one of his shifts in the library that Fischer decided he might as well use the time to do something productive, so he started browsing through maths journals. He found an interesting geometric setting whose symmetries produced well known sets of groups. But then he made one of the classic moves in the researcher's arsenal: 'I wonder what would happen if I change one of the conditions in the geometry and look at the symmetries of this new system?'

Fischer was quite pleased to find that making such a change gave him a new geometric way of looking at the shuffle symmetries. But during a talk at the University of Warwick where he was explaining his proof that the symmetries of these new geometries had to be one of the shuffle groups, someone raised a hand and pointed out that the proof couldn't be right. Some of Lie's groups could also arise as symmetries of such geometries. It is a deadly moment when your theorem blows up in mid seminar, but Fischer could see that this person was right – he must have missed something in his proof. It was back to the drawing board.

Out of the ashes of his faulty theorem, a phoenix arose – actually, three of them. Fischer settled himself in front of the mass of coffee cups and cigarette ends that accumulated on his desk and began to look again at what sort of symmetries came out of his new geometries. There were the shuffle symmetries that Fischer had first discovered. And – as the audience member in Warwick had so devastatingly pointed out – there were also geometries that gave rise to some of Lie's examples. But in addition to these, Fischer discovered three new examples which didn't seem to belong with the others. Nor did they seem to be any of the other sporadic groups that others had constructed. But they did have something to do with Mathieu's groups – the first of the sporadic groups discovered, a hundred years earlier. Fischer's new groups contained the three biggest Mathieu groups, denoted by M_{22}, M_{23} and M_{24}. So Fischer's groups quickly became known as Fischer 22, 23 and 24. And the biggest of them, Fi_{24}, was huge: it consisted of 1,255,205,709,190,661,721,292,800 symmetries.

Fischer didn't stop there. Some years after his first discovery, he tweaked his geometries a bit more and glimpsed something that made him think that there might be three even bigger examples sitting out there that were likely to dwarf both Conway's and Fischer's groups. If they did exist, they would be much harder to identify and pin down.

After his first meeting with Conway in Oberwolfach, Fischer made frequent trips to Cambridge to discuss his ideas with Conway. At one of these sessions at Cambridge, Conway decided that Fischer's three hypothetical symmetry groups needed names, because they were constantly confusing which one they were talking about. 'Let's call the smallest of the three Baby Monster, the second Middle Monster and the biggest Super Monster.' Fischer liked the idea of a Baby Monster. In Germany there was a cartoon character with this name. Given the estimated size of the symmetrical objects, 'Monsters' seemed very appropriate. Some mathematical trickery later revealed that the Super Monster was going to be impossible to build: there were certain features that contradicted each other. It was just a mirage, which vanished on closer scrutiny. But the other two were still looking robust. The Middle Monster was rechristened simply the Monster.

The first task was to pin down exactly how many symmetries these hypothetical objects might have. Just before the annual meeting at the retreat in the Black Forest, Fischer established the size of the Baby Monster (if it existed). The number of symmetries would be a staggering

4,154,781,481,226,426,191,177,580,544,000,000

The news caused a huge buzz among the conference participants. One senior member of the community who would have been really interested in the discovery was absent that year, in Australia. Graham Higman was the mathematician at whose seminar Conway had introduced the world to Conway 1, 2 and 3. 'Let's send him a letter telling him about Fischer's calculation,' someone suggested. Then one of the Oxford mathematicians, who knew Higman well, said, 'If you want him to read it, then just send him a postcard with the size of the group and nothing else.' So a postcard was dispatched from the Black Forest to Australia with a single number on its back: the number of symmetries in the Baby Monster.

Shortly after the meeting, Fischer visited Conway again in Cambridge. After various discussions, they now felt in a position to attempt to calculate how many symmetries there might be in the huge group, the Monster, that seemed to be hovering out there in the mathematical mists. They realized that its size meant they were going to need a calculator of some sort to do the computation. This was the early days of computing machines, but Conway said that he had a machine back at his house that should do the job.

The trouble was that Conway's daughters – four of them by now – had taken the machine to pieces, and there were parts scattered all over the house. Conway and Fischer spent the evening gathering all the parts together so they could rebuild the machine and make their calculations. But their stomachs overtook their appetite for large numbers, and they went out to dinner instead. Fischer received a letter a few days after his return to his new university, at Bielefeld. Conway had reassembled the calculating machine and had got his girls to calculate the size of the Monster. If it existed, it was going to have a colossal

808,017,424,794,512,875,886,459,904,961,710,757,005,754,368,000,
000,000

symmetries. But it still wasn't clear whether it did exist.

Conway and Fischer were halfway there. Their radar had picked up the Monster and they had identified how many symmetries it had. The quest was now on to see whether there really was something there or whether it would disappear, wraith-like, as the Super Monster had. It was not going to be an easy task because the calculations to establish the Monster's existence were well beyond the reach of the computers that were available in the early 1970s.

Fischer, though, was finding it increasingly difficult to concentrate on mathematics. The new university of Bielefeld that he had joined was being engulfed by radical student politics shipped in from Berlin. The Maoist movement decided that group theory was a reactionary subject of the old regime, and started protesting at the increasing number of professors in the subject being appointed. Demonstrations erupted outside the maths department with protesters holding placards demanding 'No more group theory'. One new appointee in group theory was frightened off and took a job elsewhere. During one demonstration, the students scaled the outside of the building and scrawled 'Group Theory Department' on the wall. One member of the Maoist group was Fritz, my collaborator in Bonn. He was eventually expelled from the movement after being accused of collaborating with group theorists in the department. Despite his political ideals, Fritz couldn't resist discussing mathematics with his professors.

With the noise of demonstrators outside his window, it was hard for Fischer to think about constructing a symmetrical object with more symmetries than there were atoms in the sun. But even in the peace and quiet of Cambridge, Conway wasn't making much headway either. They both believed that somewhere out there in the far reaches of the murky mathematical universe, a Monster was waiting for anyone brave enough to go looking for it.

14 June, Stoke Newington

I sometimes wonder whether it is the gradual increase in academic politics and administrative and family distractions that stifles a mathematician's creativity past the age of 40. It's like a gradual crescendo of noise outside the window which makes it harder and harder to achieve the meditative state necessary for mathematical inspiration. I often tell my graduate students to make the most of the lull they are in before the storm of life strikes them on the other side of their doctorates. This month, I seem to have spent my time doing anything but mathematics.

It's been one long round of meetings: meetings to discuss the construction of a wonderful new department of mathematics in Oxford; meetings to find ways to encourage more young people to join us on our mathematical crusades; meetings with politicians to explain why mathematics is fundamental to the technological and economic well-being of the country; meetings about the running of my college in Oxford; a meeting in Germany to deliver a seminar on my work.

There are reports to write: referee's reports for journals on the quality and correctness of my peers' research, a process that can take a very long time; reports on grant applications to fund conferences and young post-docs; letters of reference for my students to support them in their next steps.

And there is also a clutch of strange requests I can't resist helping out with: doing some mathematical theatre workshops for my favourite theatre company, Complicite, who are preparing a theatre piece about mathematics and Ramanujan; recording the voice-over for a radio programme about the French mathematician Marin Mersenne; providing Talk Sport Radio with equations to explain why Wayne Rooney is such a good footballer.

Or perhaps these are really all excuses for my lack of productivity this month, and I should just face up to the fact that my self-discipline was not up to resisting those afternoon clashes between the likes of Togo and Korea during this year's World Cup.

But in June there is also the anniversary of a phone call which caused a major and very welcome distraction. Two siblings had just arrived at an orphanage in Guatemala. We were the only family on their books approved to adopt more than one child. But we needed to make a decision there and then. Did we want to adopt these two children?

July: Reflections

Symmetry is a characteristic of the human mind.
ALEXANDER PUSHKIN, letter to Prince Vyazemsky,
25 June 1825

As more and more sporadic groups of symmetries kept sprouting up in the early 1970s, mathematicians began to wonder where it might all end ... or whether it would ever end. Conway and Thompson were divided on the issue. One thought that there were infinitely many of these exceptional groups of symmetries; the other thought there were a finite but very large number of them. Like Ancient Greek philosophers arguing over whether matter was infinitely divisible or atomic, they debated whether these groups would go on for ever or eventually run dry. Six months later, they were still arguing but had reversed their opinions. It would turn out that both of them were wrong.

As the number of interesting and varied sporadic groups increased, Conway decided to embark on a mission to chart the contours of these strange symmetries. He had alongside him a research student, Rob Curtis, who had proved his mettle exploring the contours of Conway's groups. Together they applied for a grant to put together an Atlas, as they dubbed it, which would contain everything they knew about these strange sporadic groups, together with the shuffle groups and the Lie groups.

The building blocks of symmetry that had been discovered since Galois's breakthroughs were like the hydrogen and oxygen of the world of symmetry. Groups such as the Monster, if it existed, were like heavy atoms of uranium or plutonium. But Conway and Curtis's plan was for something more than just a periodic table of symmetry. Their document was going to record all the nooks and crannies, the inlets and mountains of these symmetrical objects. This was going to be an Atlas whose pages would chart precise contours, allowing others to navigate their way through the world of symmetry.

For Conway and Curtis, the 16 types of Lie group and the shuffle symmetries of a pack of cards were like continents in this world. Each continent contained an infinite number of possible groups of symmetries. For example, the shuffles of a pack of cards required you to say how many cards were in the pack. Specifying the number would be like focusing on a particular country in the continent. The smallest country was the group of even shuffles of five cards. This symmetry group, discovered by Galois, would become the first entry in their Atlas. But for Conway and Curtis, it was the strange isolated islands – the sporadic symmetry groups – that were the most interesting. They wondered just how many of these islands, sitting on their own in the ocean, not part of any of the huge continents, lay waiting to be discovered by the symmetry searchers. Conway had found three. Curtis, a young research mathematician, was desperate to get his name attached to something. In the meantime, he would help Conway prepare the Atlas of the known world of symmetry.

Conway and Curtis got their grant and went down to Heffers, the local bookshop in Cambridge, where they purchased a huge ledger costing £80 – a considerable sum in the mid 1970s. The ledger was designed to expand as new pages were added. At this stage no one knew quite how big the book would have to be. If there were infinitely many of these sporadic islands, it would have to be an infinite book, something that would probably have appealed to Borges.

The grant also paid for an office in the department which became the Atlas headquarters. Conway and Curtis would work their way through journals, gathering as much information about these groups as they could find. When all their notes were ready for a particular group they would 'Atlantize' the group of symmetries. This would involve transferring the information onto large sheets of Atlantic blue paper. It was as if they were mapping this island of symmetry in the middle of the ocean of the page.

The mess inside the Atlas office increased as the Atlas was built, and Conway started referring to it as Atlantis: 'everything was starting to sink without trace under the mess'. Along with the mess they were also having to cope with uninvited visits by one of the department's research students. This was Simon Norton, the strange mathematician I had gone to visit on my first trip to Cambridge at the beginning of my own research career.

Norton had excelled in mathematics from a very early age. His precocious talents had earned him a scholarship to Eton. He had performed so well in mathematics that his teachers enrolled him for a mathematics degree at London University during his school years. When he arrived at Cambridge he went straight into the final year of the undergraduate course, which he came top in. But his superior mathematical talents were matched by distinctly inferior social skills. Apparently, the Norton I had met on my visit was a distinctly more socially adept version than the one that arrived in Cambridge as a student.

At Eton, the other children had soon found out what his weakness was and exploited it mercilessly. He hated being touched. His classmates would revel in laying a hand on him, then running off as Norton gave chase in an attempt to tag his assailant and somehow purge himself of the physical contact that had been made. At Cambridge the students were slightly more tolerant; mathematics departments are safe havens for many weird and wonderful characters. But he was still rather spitefully referred to as 'the child' by other graduates. Indeed, some of the students were rather disappointed to discover that Norton had been born on 28 February in a leap year. A day later, and in true Gilbert and Sullivan fashion he would have remained a child for ever. He had a strange habit of pacing the department mumbling 'ooze' under his breath until he had cracked a particular problem.

Conway was beginning to feel rather irritated by this young man who sloped into the office and sat there while they were working, every now and again making comments on their progress. Invariably there would be a fleck of spittle at the corner of his mouth, and he never seemed to change his clothes, which were accumulating an increasing number of holes. It wasn't that he was poor – his parents ran a jeweller's in London's New Bond Street – he just didn't care about his appearance. Whenever he talked he would shake uncontrollably. After a few of Norton's visits, Conway took Curtis aside: 'I can't bear it any longer. If that chap keeps coming into the office I'm just going to give up the whole project.' But within a few weeks Conway had changed his mind. He saw how invaluable Norton was becoming to their attempts to survey the landscape of symmetry. He had an exceptional talent for detail and very soon was talking about the intricate workings of these sporadic groups as if they were old friends. The sheer immensity of the calculations that were involved in charting each symmetry group's characteristics did not faze him at all. Norton was welcomed on board.

The team was quite a contrast of personalities. Conway was like a wonderful magician, inventing games and tricks to entertain the students in the department's common room. Philosopher's Football, or Phutball as it became known, was a game played with go pieces that was invented in the common room. But every game had a serious mathematical angle to it. Conway's analysis of the real game of go led to a whole new class of numbers, called surreal numbers. Curtis, by contrast, was rather dapper, tall and handsome, like an army officer who had just stepped out of a Jane Austen novel. He had, according to some, a reputation in Cambridge circles as a womanizer. Norton did not strike much of a chord with the ladies. His parents tried desperately to get him married off, and arranged a partner for him for the Trinity May Ball. It must have been a disastrous blind date, given that Norton's favourite topic of conversation was bus timetables.

Despite their different characters, they made a brilliant team. The Atlas kept expanding. Conway would carry it round the department on his shoulder, like a hod full of bricks. Eventually it became so large that it exploded on them when they tried to add another page of Atlantic blue: 'We tried to close the thing, and ... bang! ... it burst. We'd finally added the straw that broke the camel's back.' Conway always relished a challenge, and the solution proved to be right outside their door. The common room was full of decrepit old chairs whose fake leather covering had also burst open, revealing the stuffing inside. Conway simply cut off a length of this fake leather and, with the help of a huge shoemaker's needle, bound the Atlas together again.

As well as charting the contours of existing groups, the team were also finding new ones. Conway remembers 1974 as one of the most exciting periods in his mathematical life. The Cambridge team were joined for a year by a Japanese mathematician, Koichiro Harada. Fischer would often drop in. Conway remembers a table in the common room being manned by mathematicians 24 hours a day for several weeks. Harada, in true Japanese style, would remove his shoes and sit cross-legged by the low table in the common room where these groups were being forged. The others would sprawl in the ancient chairs, working furiously. The great Thompson would pass by the table in the morning to see how they were getting along.

Around about this time, mathematicians began to realize that the end might, amazingly, be in sight. This was prompted by activity on the other side of the Atlantic. Daniel Gorenstein had begun to mobilize forces to establish the limits of this world of symmetry. In 1972, in an operation of almost military precision, he had drawn up a 16-point plan to explain how to construct a complete list of all the building blocks of symmetry. At first, Gorenstein found it difficult to enlist people to his cause: 'The programme was met with considerable scepticism. I doubt that I made any converts at the time. The pessimists were still strongly in the ascendancy.'

However, within a few years people were beginning to feel far more positive about Gorenstein's programme, and began signing up to help out in what he christened the Thirty Years War. His office became like an army headquarters, with the telephone ringing and orders being given out. People were assigned different parts of the operation and were dispatched on their own missions. One of the key captains in Gorenstein's offensive was Michael Aschbacher, whose determination and mathematical abilities drove the assault at a much faster pace than might have been expected. The Americans were soon closing in on the team at Cambridge, who were scouring the world of symmetry for new groups before it ran dry.

By the mid 1970s, a total of 25 different sporadic groups had been discovered or conjectured to exist. Mathematicians could really feel the tide beginning to turn – the feeling was that 25 might be the limit of what was possible. Then the man who had started the whole thing off in 1965, with the first sporadic group since the five of Mathieu, announced the possible existence of a 26th group. It was the fourth that Zvonimir Janko had discovered. Like two bookends, Janko 1 and Janko 4 seemed to represent the beginning and the end of the exploration of this strange archipelago.

Moonshine

By 1978, 24 out of these 26 symmetry groups had been built. There were just two that remained out there in the mists: the Monster and Janko's fourth group. The evidence for the existence of this pair was very convincing, but mathematicians still had the task of constructing something whose symmetries matched up with the numbers that were being predicted.

Conway's Atlas now contained extensive information about the Monster, even though the Cambridge team still weren't sure that it really existed. One useful piece of information was the discovery that the smallest number of dimensions in which anyone could construct an object whose symmetries were those of the Monster would be 196,883. This was why the object deserved to be called the Monster. There are Lie groups which have more symmetries than the Monster was predicted to have, but they were symmetries of a geometry that could be constructed in eight dimensions. But you had to travel into 196,883-dimensional space before this Monster would show its colours. The reason people felt that the Monster was so out of reach was that manipulating an object in 196,883-dimensional space was well beyond the capabilities of even the most powerful computers available in the late 1970s.

As Conway was about to learn, the number 196,883 would reveal the Monster to be not just some bizarre entity at the very fringes of mathematics, but an object connected to some of the most central and striking parts of mathematics. And again, the messenger was his old friend John McKay. The message was sent in a letter addressed to Thompson. It announced the discovery of this equation:

$$1 + 196{,}883 = 196{,}884$$

You might think that McKay had lost his mind. Surely anyone could write down any number of such apparently trivial identities. But for McKay, the numbers in this equation weren't just random.

Sitting in his office in Concordia University in Montreal, McKay had been idly browsing through a paper on a subject completely unrelated to group theory. The paper was about one of the central and most mysterious objects in number theory, called the modular function. In the 1990s, variants of the modular function would turn out to be the key to solving Fermat's Last Theorem. By this time, 1978, the modular function already had a rich heritage going back to the nineteenth century. Felix Klein, who had been Lie's partner when they explored the connections between geometry and symmetry, had investigated the object and had even used it in showing how to solve the quintic equation, using more complicated tools than simply taking roots of numbers.

The construction of this object depends on a sequence of numbers which starts 1, 744, 196,884, 21,493,760, 864,299,970, . . . Essentially it is an infinite polynomial:

$$x^{-1} + 744 + 196,884x + 21,493,760x^2 + 864,299,970x^3 + \ldots$$

and the sequence of numbers tells you the multiples of x, x^2, x^3, . . . as the polynomial spirals off to infinity. McKay looked at this sequence of numbers and thought that one of them looked awfully familiar. He pulled out some papers about the shape of the Monster that everyone thought must exist – and there it was. Well, almost. 196,883, the smallest dimension in which you would be able to see the Monster was just 1 less than the third number in the sequence for the modular function. McKay could have dismissed as pure coincidence the fact that the numbers were so close, but he happened to believe that mathematics was full of such special connections. So he decided to record his 'discovery' in a letter to Thompson.

The simple equation in the letter raised an interesting question: was it just a coincidence that the numbers were so close, or was there some deep hidden connection between the Monster and number theory? Initially, Thompson dismissed the connection, comparing it to reading tea leaves. After all, one has to be very careful with numerology. For example, in the sixteenth century Kepler used some startlingly strange coincidences in number patterns to explain why the six known planets were inextricably linked with the five Platonic solids. The discovery of a seventh planet blew Kepler's numerological coincidence out of the water, although Kepler had already abandoned his theory for other reasons. But when Thompson started to play with some of the other numbers in the modular sequence, he found that the Monster was even more interwoven with this object from number theory.

Thompson looked up the next dimension in which people were expecting to see the Monster appear. It was quite a shift up to 21,296,876. At first sight, things didn't seem to match up. Then Thompson spotted how to fit this number into a pattern connected to the numbers for the modular function:

$$21,493,760 = 1 + 196,883 + 21,296,876$$

This was beginning to get spooky. He tried the next number up. The next-highest dimension in which you can see the Monster is 842,609,326. Again, the pattern seemed to break down: Thompson couldn't get the next number in the modular sequence, 864,299,970, simply by adding up the dimensions of the Monster. But then he spotted a trick that would make all the numbers match up. It looked slightly contrived, but the fact that it worked surely meant that there was more to all this than mere coincidence. He could get 864,299,970 by adding up the following numbers, all numbers related to the Monster:

$$1 + 196,883 + 1 + 196,883 + 21,296,876 + 842,609,326$$

McKay had also gone on to find these extra equations, but it was Thompson who first published them. McKay admits that 'I was a bit peeved really. I don't think Thompson quite knew how much I knew.' Thompson was not convinced that the equations had any deeper meaning, and still referred to them as numerology, a word generally associated in mathematics with quacks. Thompson had received McKay's letter while he was visiting Princeton. As soon as he got back to Cambridge he showed McKay's equation to Conway.

Conway, though, was in possession of secret information about the Monster unavailable at that time to anyone outside the Atlas office. Indeed, Conway believed that the Atlas they were compiling was becoming a book with all the answers in it. He wasn't even sure whether publishing it would be such a good idea, for that would give away the advantage that his team in Cambridge now possessed. The

Atlantic blue pages of the Atlas now contained complete details of
what the contours of the Monster would look like (if it was ever
located).

As soon as Thompson showed him McKay's 'numerology', Conway
sped down to the library to check in a book on number theory which
had a complete description of the modular sequence. Before long he'd
started to build equations which laced the numbers of the Monster
together with the sequence of numbers for the modular function. His
mathematical partner in crime, Norton, was away on a two-week
journey around the rail networks of England. By the time he got back
to Cambridge, Conway had made good headway. Once Norton had
got these numbers between his teeth, he soon had a complete set of
equations that showed how to go from the set of numbers in the
Monster to the sequence of numbers in the modular function. 'Thank
God I had that two-week head start on Simon,' mused Conway, 'else
I wouldn't have got a look-in!'

Conway christened the extraordinary numerology they had concocted out of McKay's simple equations 'moonshine'. Although they
had inextricably laced these two things together, no one knew what it
all meant. And there was still the problem that no one had actually
built the Monster. But then came another startling piece of news
from across the water. The Monster was no figment of mathematical
imaginations – someone had finally given flesh to the bare-bones
description in Conway's Atlas.

The mathematical Doctor Frankenstein

Bob Griess had been on the trail of the Monster since the early 1970s.
In 1973 he had made the same calculations as Fischer that led to the
prediction of this huge symmetrical object, but frustratingly, he'd seen
Fischer get most of the credit. A year earlier Griess had been chasing
the construction of another group, predicted by Arunas Rudvalis, but
was pipped to the post by Conway and a colleague, David Wales.
Conway was so aware of the race to build this group that he had
recorded the time and date of the group's delivery: 4 p.m. on 3 June.
Just before the hour struck, the phone went in Conway's flat in Caltech,
where he was visiting. He knew it would be the computing lab with

the results of their construction. He waited for the clock to count down to 4 o'clock exactly . . . 4 seconds, 3 seconds, 2 seconds, 1 second . . . then picked up the receiver. 'You are the proud father of a new group of size 145,926,144,000,' announced the caller from the lab. The group had been predicted by Rudvalis a month before, at 3 p.m. on 4 May.

Understandably, Griess was keen to get his name on at least one new sporadic group. The problem was that they were clearly beginning to run dry. Gorenstein's classification programme was starting to close in, and it seemed that only two of the 26 islands were still unclaimed. But going out and finding the Monster was easier said than done. No one seemed to have the tools to accomplish the massive task. As Griess points out now, 'How can you try to demolish Everest if all you've got is a toothpick?' However, news of the evidence connecting the Monster with the modular function from number theory spurred him on. It showed that the Monster wasn't a strange quirk of nature. If it existed, it was going to be connected to some of the most important bits of mathematics.

By 1979, Griess felt brave enough to try to tackle this object which existed in 196,883-dimensional space. He had a year's sabbatical at the Institute for Advanced Study at Princeton – a perfect setting for mounting his assault. He started to work around the clock on his construction. There were now lots of clues about where to look for this missing symmetry group. 'The whole thing felt like a detective story. The trouble was that you didn't know if you were going to find treasure or find that you'd wasted your time up a blind alley.'

Griess found night-time the most productive. He got to know the cycle of the security guard who would do his rounds, passing by his office every two hours. He came across strange individuals who would camp out in their offices at night rather than pay for accommodation. And gradually he started to see the Monster emerging from the mists.

As December 1979 arrived, Griess was working harder and harder. 'I was determined to beat the change of decade.' He failed, but only by 14 days. Having taken just one day off, for Christmas, by 14 January 1980 he'd discovered an object whose symmetries matched up with the predictions that everyone was making for the Monster. It was an extraordinary tour de force. Bare-handed, without recourse to a computer, Griess had constructed this enormous object which possessed more symmetries than there are atoms in the sun.

Griess now faced the daunting task of writing up his discovery, but having been burnt several times before he sent out an announcement to make sure that this time he got his name on it. He was also rather upset that the group had already been christened the Monster, whereas all the other groups, apart from the Baby Monster, were named after the people who had discovered or constructed them: 'The fact that it was called the Monster hurt me. I was rather unhappy about it.' So in his announcement he also tried to get it renamed. He knew that he wouldn't be able to convince the mathematical community to abandon the colourful name in favour of the Fischer–Griess group. So he went instead for calling it the Friendly Giant. To see the F and the G there to recognize his and Fischer's involvement would suffice:

> For me, a Monster smacks of evil dictators and my work was a taming
> of this object. So I felt my name was also a serious statement, that this
> was a friendly accessible object. Monster is a name that will scare all
> but the bravest to dare look at it. I think things should be more open. Let
> us tame nature and make it more understandable. That's my attitude.

The name never caught on.

After the initial excitement when Griess announced his construction of the Monster in early 1980, mathematicians grew exasperated at the complete lack of details of the construction. The Cambridge team were especially frustrated, and John Thompson, who had been Griess's supervisor in the early 1970s, made a special trip across the Atlantic to find out exactly what Griess had done.

A week later, Thompson arrived back in Cambridge. Conway eagerly buttonholed him in the common room. 'So, what's the construction all about?' Thompson looked a bit sheepish. 'Gee, well I guess he didn't mention it.' This was extraordinary. The world's greatest group theorist had flown across the Atlantic to learn about one of the greatest achievements of group theory in the twentieth century – and his student hadn't even mentioned it? Conway looked incredulously at Thompson: 'Didn't you ask?' 'Well, gee, I guess I didn't think it was my place.' Griess seemed determined to keep things close to his chest until he had written down all the details for publication.

A month after Griess's announcement, Conway and his team managed to build the 26th sporadic group, the last one that mathematicians believed existed. In contrast to Griess's bare-hands construction of the Monster, the fourth Janko group relied on hours of computing time for its entrance onto the mathematical stage. Conway had by this time enlisted a fourth member to the Atlas team who was a whiz with the computer.

As far as the academic establishment was concerned, Richard Parker was something of an outsider, but he loved doing maths. After getting a first in mathematics at Cambridge, he left to earn his money writing software for cash registers. In 1978 he met up with Conway in Cambridge. Conway was desperate for some automation of the tables they were producing for the Atlas. Parker was keen to complement the lucrative job of programming tills with something a bit more meaningful. The match was perfect.

Parker had been interested in mathematics from an early age. Indeed as a teenager he used to hide in his room solving maths problems for such long periods that his mother became worried by his solitary, sedentary habits, so much so that she dragged him off to their local doctor. 'There is something wrong with my son', she complained, 'he sits in his room all day studying mathematics.' The Doctor replied unhelpfully 'I wish my son did mathematics all day'. It was the local badminton club that came to the rescue. Parker was someone who never did things by halves. By the time he joined the Atlas team he was playing badminton for club and country.

The Atlas team were always cooking up crazy schemes. Once they took out an advertisement in *The Times*: 'Gentleman, thinking of starting new religion, seeks converts.' Any respondents wouldn't have known, of course, that the religion required dedicating oneself to a Monster and chanting endless numbers in the moonshine.

The Atlas work soon sucked Parker in, and he decided to go freelance to make more time to dedicate to Conway's project. The greatest challenge was getting the computer to build Janko's fourth group. It required working in 112 dimensions, a lot smaller than the Monster but still something that stretched the limits of Parker's computers. By February 1980 they'd found a way to check whether the symmetries of this geometry in 112 dimensions really did give this final sporadic group. As the computer churned away, returning the answer 'yes' to the millions of questions it was being asked, a sense of anticlimax descended over the team. They knew that this was probably going to be the last of the sporadic groups. They could predict the time of the final 'yes', and when it came Conway recorded the time to the second and wrote it down on a big piece of card with the words 'J_4 constructed'. He pinned it wistfully to the Atlas office door. That was probably it.

By the summer, Griess at last felt ready to go public with the details of his construction of the Monster. He started giving talks. The seminars drew large audiences of mathematicians eager at last to find out how he'd done it. The crew in Cambridge finally got hold of a copy of the 100-page paper that Griess had been preparing and pored over the details they had been unable to construct for themselves or that Thompson failed to elicit from his former student. News started to circulate that Gorenstein's team were also in sight of their goal of proving that the 26 groups, the prime-sided polygons, the shuffle groups and the Lie groups were all that would be in the Atlas.

No one can quite date the end of the journey. In 1980 *The Mathematical Intelligencer* was going to publish its fourth issue of volume 2 with a cover carrying the words 'the 26 known sporadic groups'. But in proof the 'known' was struck out and a note added announcing that 'the classification of finite simple groups is complete. There are no more sporadic groups.' Gorenstein too was declaring that it was 'all over' in 1980, but later amended his dating of the end to February 1981, when a paper of Norton's proved that there couldn't be two different groups that looked like the Monster. Without Norton's contribution there was a theoretical possibility that there might be several sporadic groups with the same number of symmetries as the Monster. Others opted for 1983, when certain important pieces in the jigsaw finally fell into place.

By 1985 Conway, Curtis, Norton and Parker, together with a fifth author, Rob Wilson, who joined their endeavour, were ready to submit the *Atlas of Finite Groups* to the publishers. The journey was over. The proof was complete. By the time I arrived in Cambridge, the Atlas – a record of a two-thousand-year journey through symmetry – was running off the presses.

5 July, Edinburgh

Two years ago to the day, I attended a conference in Edinburgh to celebrate the 25th anniversary of the discovery of moonshine in the Monster. It was a reunion of many of those who had been involved in the journey. Even for mathematicians, this was a fairly weird bunch of characters: Simon Norton, with bulging plastic bags full of timetables and a shirt full of holes; John McKay, looking the spitting image of Kentucky Fried Chicken's Colonel Sanders with his red cheeks and white moustache; John Conway, now less hairy than in his Cambridge days, though still with a wild glint in his eye and a toothy grin. The secretary handing out the conference material at the registration desk was looking quite shell-shocked. 'They certainly are a strange bunch,' she commented.

On the first day of the conference, John McKay, the first person to see the moonshine glinting on the Monster, stood up to introduce Conway: 'John will be explaining my construction of the Monster.' 'No, I'll be explaining *my* construction of the Monster,' Conway countered rather tetchily. Since Griess's first paper, others have tried to find more efficient ways to build the Monster. The hope was that each new construction might explain the moonshine and the Monster's links to number theory.

'You'd better all applaud now, before I give the talk, because it's going to be a failure.' Conway started clapping, and the audience, looking rather bemused, joined in. 'I've given this talk in Princeton recently and it was a failure there. It was a failure again in Rutgers. But I believe in try, try, try again.' He launched into his construction of the Monster, a creation he has honed ever since he received Griess's long paper. To Conway, the Monster is 'astonishingly simple' and he is desperate to communicate his vision.

Just as Conway had predicted, though, within about ten minutes most of the audience looked quite glazed. Several people, including McKay and the man next to me, had fallen asleep. McKay's snoring began to get rather distracting, so someone dug him in the ribs to get him to stop. Halfway through, Conway could feel that he'd lost his audience. 'I told you it would be a failure. But you've clapped already.' You could tell that he was deeply frustrated. 'The group's big but it really is rather trivial.' When you've lived with the group for nearly thirty years it does become a friend rather than a Monster.

After the lecture, Conway was downcast. We sat around trying to dissect what had gone wrong. 'There's this wonderful quote from Shakespeare's *Henry IV Part One* that I try to live up to,' he said. 'Owen Glendower declares: "I can call spirits from the vasty deep." But then Hotspur counters, "Why, so can I, or so can any man; But will they come when you do call for them?" I want to be like Hotspur.' That morning, though, the spirits didn't come. The Monster stayed in the vasty deep.

Conway has this insatiable need to be able to summon knowledge from the depths of his mind. That's why he learnt π to so many decimal places when he was younger. He also taught himself how to work out the day of the week for any date you care to give him. Usually this is a skill that is associated with autistic savants, but Conway has taught himself the mathematics behind it. To keep his mind active, he has set up his computer so that instead of entering a password to gain access to it, he has to identify the day of the week for ten different dates within 12 seconds. He believes that by now he is the fastest in the world at doing this, and that no one else can hack into his computer. He also believes that he can teach the trick to anyone – although perhaps not at the lightning speed that he has achieved.

At the conference dinner that evening, Conway explained the technique to one of the young female graduates. As we all walked back to the university from the centre of town, we passed through an old graveyard. Here was the perfect opportunity to put the graduate through her paces. Conway stopped at a gravestone and started to read the inscription. 'OK, Alexander Maclean, Perfumer to Edinburgh. Died 5th October 1834. What's the Doomsday for that year?' After only a short pause, back came the answer that Maclean died on a Sunday.

Conway was quickly onto his next chat-up line. The sky was full of stars, with a few clouds scudding across the firmament. 'I can name every star in the sky,' he boasted. 'You see that cloud there? I can even tell you the names of the stars behind that cloud.' And he started reeling them off: 'Betelgeuse, Bellatrix, Alnitak, . . .' A young graduate joined in and showed off his knowledge too. It was very curious. To me, this sort of knowledge seems quite pointless. It is just butterfly collecting, with no sense to it. It was just the sort of thing I moved into mathematics to avoid.

In 1982, Thompson similarly described the classification of the building blocks of symmetry as 'an exercise in taxonomy. To be sure, the exercise is of colossal length, but length is a concomitant of taxonomy.' Ever since the classification was finished, mathematicians have wondered whether there is a more conceptual explanation of why the Atlas contains the building blocks it does. For those addicted to the pursuit of patterns, these 26 sporadic groups just don't make sense. They are just sitting there like a random constellation in the night sky. Thompson went on to write:

> Not surprisingly, I wonder if a future Darwin will conceptualize and unify our hard won theorems. The great sticking point, though there are several, concerns the sporadic groups. I find it aesthetically repugnant to accept that these groups are mere anomalies . . . Possibly . . . *The Origin of Groups* remains to be written.

The sheer length of the proof of the classification, which covers some ten thousand pages in the mathematical journals, has been a concern. Can we really be sure that we've covered every base? A proof of this huge extent is 100 per cent guaranteed to contain errors – but are any of them fatal errors? Some mathematicians take care to state very clearly when proving a theorem if they are using the classification of these building blocks at some point in the argument. It's almost quoted as a working hypothesis rather than a proven theorem just in case it turns out that there is a 27th sporadic group that was missed from the list and the proof needs to be re-evaluated.

Indeed, in the late 1980s, news spread that one piece of the jigsaw was actually still missing. Geoffrey Mason had been responsible for writing a paper to cover one of the difficult steps in Gorenstein's 16-point plan to complete the classification. Gorenstein had originally assigned this project to Janko, but the Croatian had walked away from it in 1975, after five years of investigation, declaring it too difficult. One member of Gorenstein's regiment wrote that Janko's resignation was the only time he saw Gorenstein's resolve falter. Mason had taken up the challenge abandoned by Janko and had produced an 800-page preprint with a proof of the missing step, but it had never been published. When those involved in writing up a complete account of the whole proof of the classification started to look at Mason's paper in 1989, they understood why. It was full of holes.

The party to celebrate the completion of the classification had been somewhat premature. As one mathematician close to the proof admitted,

> Viewed retrospectively and soberly, it was perhaps a bit hasty to claim that everything was finished before the manuscripts had been checked carefully, but it was quite understandable. Mathematics is done by human beings who have an emotional aspect to their personalities in addition to a rational one.

Criticism began to mount. How could you call this a theorem if there was an 800-page hole in the middle of it? One of the great mathematicians of the twentieth century, Jean-Pierre Serre, was particularly critical:

> For years I have been arguing with group theorists who claimed that the 'Classification Theorem' was a 'theorem', i.e. had been proved. It had indeed been announced as such in 1980 by Gorenstein, but it was found later that there was a gap. Whenever I asked the specialists, they replied something like: 'Oh no, it is not a gap; it is just something which has not been written, but there is an incomplete unpublished 800-page manuscript on it.' For me, it was just the same as a 'gap', and I could not understand why it was not acknowledged as such.

Gorenstein died in 1992 with the controversy still raging. Three years later Gorenstein's captain, Aschbacher, in collaboration with Stephen Smith, set about filling the gap. In 2004, the same year as the Moonshine conference in Edinburgh, a plug for the gap in the proof was published. Mathematicians had thought that the 255 pages that made up Thompson and Feit's paper in the early 1960s was big. Aschbacher and Smith's paper dwarfed it: it ran to 1,221 pages. There is a general consensus that even if there are more errors or gaps still remaining, they will not be fatal. Smith believes that 'the basis for the reliability of the proof is that very many parts of it are extremely parallel. It's very unlikely there's a hole you could drive a truck through.' In other words, there are so many threads woven into this proof that pulling one won't make the whole thing fall apart.

With Gorenstein's death came the realization that the mathematicians who truly understood all the intricacies of the proof of the classification were getting old. Now that the proof was complete, few young and aspiring mathematicians were attracted to the field. The classification involved very specific techniques that people are now beginning to think could die out with the passing of this generation of practitioners, almost like the craft and skills of the medieval stonemasons, which have never been replicated. All the more reason for those engaged in rationalizing the proof to make sure that they haven't missed a 27th group.

Conway et al. freely admit to the possibility of errors in their Atlas. In the introduction they write that

> with regard to errors in general, whether falling under the denomination of mental, typographical, or accidental, we are conscious of being able to point to a greater number than any critic whatever. Men who are acquainted with the innumerable difficulties attending the execution of a work of such an extensive nature will make proper allowances. To these we appeal, and shall rest satisfied with the judgement they pronounce.

These are not actually Conway's words, but are taken from the preface to the first edition of the *Encyclopaedia Britannica*, published in 1771.

In fact the first error in the Atlas is a glaring typo at the beginning of the introduction, where the first heading is 'Prelimaries'. Such errors are of course harmless. But is it possible that the Atlas has missed a 27th sporadic group sitting out there somewhere in the ocean of symmetry? Perhaps someone looked out in one direction and thought that the maths told them there was nothing to find there. It very nearly happened with a number of the groups that we did discover.

Conway recalls that the Rudvalis group that he constructed at 4 p.m. on 3 June had at one point thrown up a contradiction the previous month that should have made them turn back:

> The contradiction refused to go away even after we had condensed it onto one side of a sheet of paper and scrutinized it for several days.

Fortunately we were so convinced that the group existed that eventually we just put that piece of paper aside and constructed the group by another method that carefully went nowhere near the contradiction! What worries me is the nagging thought that another group like the Rudvalis group might have been disproved somewhere in the classification programme by someone who had no overwhelming conviction that it existed.

Conway almost found something that everyone had missed: 'I remember great excitement one night when we thought we'd found a 27th group.' You could see the thrill in Conway's eyes as he contemplated the possibility. But it eventually turned out to be one of the Lie groups in disguise. Indeed, the likes of Conway and Fischer would dearly love there to be a 27th group that we've all missed. For these mathematicians, the groups are like jewels: the more of them, the better.

With the end of the journey and the publication of the Atlas in 1985, a sense of anticlimax descended on group theory. It had been such an exhilarating few decades that nothing could quite match that feeling of building a completely new group. Conway was offered a prestigious job at Princeton. He loved Cambridge. He loved to walk through the college gardens, weaving in and out of different colleges, thinking about mathematics. But the party was over. More serious forces were mobilizing in the department, and Conway could sense the beginning of a group theory cull. 'We were regarded as layabouts who sat around the department all day playing backgammon and go.'

The chairman of Princeton invited Conway and his family to come over for a year to try Princeton out, but Conway knew that it was just a way of delaying the decision. Eventually he suggested that they should have a coin-tossing ceremony to decide. Heads, stay in Cambridge; tails, go to Princeton. Many of Conway's decisions are made in this way. But his second wife said that was too disrespectful. Eventually, Conway went to Princeton for the challenge of something new. 'Everyone is so serious in Princeton. Everyone works so hard. In Cambridge people would just come in with crazy ideas and see where they would go. To do good work you have to be somewhat irresponsible.'

The move was not without its pressures. Shortly after arriving in

Princeton, Conway split with his wife. The separation plunged him into a deep, black depression which pushed him to attempt suicide. He awoke in hospital, quite relieved to find that he'd failed, but daunted at the prospect of stepping back into normal life with people talking behind his back. 'I thought, "What would Conway do here?" Conway would make it perfectly obvious he knew.' So he borrowed a T-shirt from a friend with SUICIDE in big letters emblazoned across the front. Suicide is the name of the second hardest rock-climb in the United States, and his friend had conquered it a few years earlier.

Conway hates getting old. For a mathematician it can often be harder than for most people. Despite our desperate desire to believe otherwise, for the great majority of us the peak of our creativity is in our youth. The mathematical menopause is a harsh reality. Another famous Cambridge alumnus, G. H. Hardy, also attempted suicide at the onset of old age. Conway stares into the future and doesn't much like what he sees: 'What's there at the end? Death, and I don't like that very much. I don't want to grow old. I don't feel old in my mind.' This is why he has started training his mind to do lightning calculations again.

Old age has also brought out a vain streak in Conway. His hair is no longer the wild mane it was in his Cambridge youth. 'After I got my hair cut I went into a local ice cream shop next door, and the girl said, "Oh, you look a lot younger." I always thought I didn't care about appearance, and I didn't care until recently.' He's now married to his third wife. She will be his last. 'Well, I had four children with my first, two with my second, one with my third . . . I'd have to have half a child with a fourth wife,' says the man obsessed with patterns. Also, the maintenance is getting quite expensive.

After Conway's move to Princeton, many others went their own separate ways. 'Once Conway had left there didn't seem any point in hanging around Cambridge,' admitted Wilson, the fifth author of the Atlas. The fourth author, Parker, continued to combine maths with shopping tills: 'I had a research grant in Cambridge from the autumn onwards, so I had more time for mathematics. This didn't really make up for the loss of John Conway.' Norton however, without the political and social skills to survive the cut and thrust of the academic world, was rather abandoned by everyone. He still went on exploring the Monster, and many mathematicians say that those plastic bags he carries around with him contain not just bus timetables but more secrets about the Monster than anyone can imagine. But they will be hard to coax out of him.

Finding moonshine

There was still one big mystery: What was it that was casting its light on both the Monster and the modular function? Conway had left behind the mystery of moonshine, but one of his PhD students who'd remained in Cambridge had caught the moonshine bug. Richard Borcherds' application to do a PhD in the department was almost rejected when friends doctored his application form and under 'sex' wrote 'yes please'. The head of department was not impressed, but Conway spotted that Borcherds was a bright student and persuaded the head of department to turn a blind eye to the cheeky comment.

Conway had been trying to prove something about the Leech lattice for about six weeks and had got completely stuck. He mentioned the problem to Borcherds. A couple of weeks later, Conway was still stuck. Borcherds looked surprised: 'Oh, you still haven't proved it? I did it last week.' Conway admits that he didn't really teach Borcherds at all – he didn't need to. Borcherds just went off and did his own thing. Borcherds has a rather Neanderthal look about him, with hair everywhere and arms that droop to the floor and seem to drag along behind him as he runs from one place to another. Communicating his ideas was not one of Borcherds' strong points. 'His first seminar was a disaster,' Conway remembers. 'After about 20 minutes he could see that he just wasn't going to present the material in time. So he just gave up and ran out.'

When Cambridge emptied out of group theorists, Borcherds was not particularly concerned. He had always been something of a loner. He liked cycling long distances by himself. His other passion was caving and potholing: being stuck underground in total darkness with hardly anywhere to move, not sure whether there was a way forward or whether you were going to able to wriggle backwards, was his idea of a fun weekend break from mathematics.

Mathematically too, Borcherds found himself squeezing further and further down a tunnel – and this one seemed to have a light gleaming at the end of it. He was beginning to believe that this was the light of the moonshine glinting off the Monster. He had become fascinated by a new and esoteric algebraic structure called a vertex operator algebra. Just as Galois's abstract idea of a group had taken decades to catch on, this structure was not yet mainstream mathematics:

> I got a bit disillusioned, because it was obvious that nobody else was really interested in it. There is no point in having an idea that is so complicated that nobody can understand it. I remember I used to give talks on vertex algebras, and usually nobody turned up. Then there was this one time when I got a really big audience. But there had been a misprint, and the title read 'Vortex algebras', not 'Vertex algebras'. The audience was made up of fluid physicists, and when they realized it was a misprint, they weren't interested either in what I had to say.

If the physicists weren't interested then, they have certainly have started listening to Borcherds now. It turns out that these algebraic structures help to underpin some of the deepest ideas of string theory, the current theory that hopes to unite relativity and quantum physics. As physicists have discovered, there is a lot of strange number theory wrapped up in string theory, including the modular function that sits on one side of the moonshine mirror. Borcherds found that these algebraic structures were also inextricably linked to the Monster's symmetries. The insight came not while he was stuck down a pothole, but on a bus journey:

> I was in Kashmir. I had been travelling around northern India, and there was one really long, tiresome bus journey which lasted about 24 hours. Then the bus had to stop because there was a landslide and we couldn't go any further. It was all pretty darn unpleasant. Anyway, I was just toying with some calculations on this bus journey and finally I found an idea which made everything work.

His calculations revealed why the numbers in the Atlas for the Monster and the modular numbers from number theory were both being illuminated by the vertex operator algebras. This connection with the algebra of string theory and the physical theory of the universe made the moonshine even more exotic than anyone had imagined. Once word got out, people started talking mystically about the Monster being 'the symmetry group of the universe'. At the very least, this strange symmetrical snowflake in 196,883-dimensional space was revealing patterns that resonated with ideas in theoretical physics.

For Borcherds, the revelation was one of most of the exciting moments of his life: 'I sometimes wonder if this is the feeling you get when you take certain drugs. I don't actually know, as I have not tested this theory of mine.' The discovery catapulted his ideas into the limelight. The audiences for his talks were now packed with mathematicians and physicists, and this time not because of a typo in the title. By 1998 Borcherds' work on moonshine was being recognized as one of the greatest achievements in mathematics. He was awarded a Fields Medal at the International Congress of Mathematicians in Berlin that year. He explains that the award did not excite him as much as the discovery on the bus in Kashmir: 'Before the award, I used to think it was terribly important, but now I realize that it's meaningless.'

The conference held in Edinburgh to celebrate 25 years since the discovery of moonshine was also dedicated to understanding the insights provided by Borcherds' work. It is curious that although his work proves the connection between number theory and this huge symmetry, there is still a sense that we haven't got to the bottom of the connection. At a recent meeting where everyone was congratulating Borcherds on his achievement, Borcherds' own supervisor, John Conway, stood up and declared: 'I come not to praise, but to bury . . .' and launched into a critique of Borcherds' work. Conway still felt that Borcherds had not truly illuminated what it was that was connecting the symmetry of the Monster with the modular function from number theory.

The mathematician Andrew Ogg had noticed some early evidence for moonshine in the 1970s and had offered a bottle of Jack Daniel's for an explanation. He recently asked Conway if he should present the bottle to Borcherds. 'No,' said Conway, 'he's proved the connection but not explained it . . .' Conway believes that a deeper explanation is waiting to be uncovered and doesn't think we should let the beast lie:

I think in two hundred years' time that someone will be looking at some natural geometric set-up and they'll see it has lots of symmetries and as they study it they'll gradually see it has the symmetries of the

Monster. Then they'll dig out all our old papers from several centuries earlier about the Monster.

During the conference in Edinburgh, a report was circulating about a piece of research which had found that mathematics departments have a higher proportion of people with Asperger's syndrome than any other university departments. A general denial rippled through the conference, but when Borcherds was interviewed for a newspaper following his award of a Fields Medal, he admitted that he thought he suffered from it: 'I've got a hell of a lot of the symptoms. I once read something in a newspaper and it said there are six signs of Asperger's syndrome, and I said to myself, "Hey, I've got five of those!"'

Hans Asperger, a Viennese paediatrician, identified the syndrome in his 1944 doctoral thesis as a high-functioning variant of autism. Some of the criteria for the syndrome that Borcherds had read about include severely impaired social interaction; all-absorbing narrow interests; imposition of routines and non-verbal communication; and in some cases clumsiness. It is also associated with a strong drive to systematize. The mathematical world seems well suited to those with such traits, and Asperger showed that many with the syndrome gravitated towards professions that exploited their mathematical skills.

Borcherds is well known among his friends for slinking off during a dinner party, to be found hours later deep in a mathematical text. He just doesn't see the point of small talk. He is not a conversationalist, can't see the point of a telephone conversation except for transferring information, and will avoid eye contact with you at all costs on the painful occasions when he does have to engage in conversation. But his traits are so typical of those in a mathematics department that his peculiarities are easily tolerated.

Simon Baron-Cohen, one of the leading researchers into Asperger's and autism based in Cambridge, read about Borcherds' self-diagnosis and was curious to interview the mathematician himself. After all, it has become quite fashionable for people to lay claim to a pinch of autism or Asperger's in recent decades. As Asperger himself wrote, 'It seems that for success in science or art a dash of autism is essential.'

Asperger's is strongly inheritable, and Baron-Cohen was not surprised to find that Borcherds' grandfather, who lived in South Africa, was the kind of man who preferred spending weeks out hunting in the bush, with no thought for his family and without missing the company of human beings. Borcherds escaped instead into mathematics, stalking alone the mysteries of the Monster. Baron-Cohen's analysis confirmed Borcherds' self-diagnosis. He presented evidence of 'extremely low empathizing, extremely high systematizing and a lot of autistic traits. His talents in mathematics have resulted in his finding a niche where he can excel (to put it mildly), and where his social oddness is tolerated.'

Queens we're in

As we were sitting around after the day's talks discussing the research on the connection between mathematics and Asperger's, Norton bustled over to join us, his bags bashing against anyone who was in his way. No one could deny that Norton displays many of the traits that Borcherds saw in himself. He'd changed his shirt once since the beginning of the conference, but the second shirt is as full of holes as the previous one. It's true that he is obsessed with numbers and train timetables. But when you dig a little deeper you discover that he is very politically active, working for the Labour Party and a key voice of Transport 2000, a political movement campaigning for public transport. While we were in Edinburgh he was interviewed on BBC Radio Cambridge about a campaign to stop the closure of a local branch line. He told me at the conference that he has travelled every stretch of the British rail network. 'Except for one,' he said, frothing slightly at the mouth at the thought of this virgin track. There is a branch line from Newquay that he is saving up for a special occasion.

Norton brandished a detailed map of the area around the campus. 'Can we go for a curry in Currie?' The discovery of the name of this little local village is too much for Norton's sense of symmetry. Conway was in two minds. 'We'll let the coin decide,' he said, taking out a 50 pence piece. 'Queens we're in; tails we're out.' It lands with the queen's head up. So dinner is in the campus cafeteria.

Conway was in an upbeat mood. His failure to summon the Monster at the beginning of the week had sat heavily on his shoulders. He'd spent the days since trying to find a better way to explain something he thought was so natural and obvious. The breakthrough was a set of pictures he'd concocted that he thought showed much more clearly something that was getting lost in the notation and equations he'd been writing on the blackboard on the first day. So he'd asked the organizers if he could give his talk again. This morning there were far fewer glazed looks, and Conway felt that he'd lived up to his desire to emulate Hotspur and raise the spirits from the deep.

We listened and watched as Conway entertained us. He had a wonderful set of card tricks that he enjoyed showing off to us. Origami frogs jumped from his hands. He even demonstrated how to make 20 coins land heads. And then all land tails. We began to wonder whether our decision to stay in that night really was down to chance. We were all sworn to secrecy not to divulge the trick behind the coins (but let me just hint that there is some strange asymmetry in many coins that Conway has found a way to exploit).

We then got a demonstration of his tongue gymnastics. He'd read as a student how 1 in 4 can roll their tongues. Out came his tongue, with a valley in the middle. 1 in 40 can make a clover. Out popped his tongue, now with three valleys. 1 in 400 can invert their tongues. Sure enough, we got to see Conway's tongue upside down. 1 in 4,000 can make their tongues go fat and thin. Out came the tongue again, now all puffed up, then deflated to a thin sliver. He explained how he'd been photographed by *Reader's Digest* after he'd responded to an appeal for volunteers who could do all four. He'd only met one other person who could match his lingual callisthenics, a woman at a party. After comparing tongues, Conway suggested that they were meant to be together, but she didn't think so and backed off quickly. 'Anyway, I don't think she could do this . . .' and Conway had set up a sine wave oscillating along the length of his tongue.

Before eating dinner, he pulled out a massive bag full of pills. 'I had a heart attack a few years ago. Now I have to eat all these with my dinner.' Always the mathematician, he has found a pattern in the colours and shapes of the pills which allows him to remember which and how many to take in the morning and in the evening.

Conway is such a good communicator, a wonderful performer, a master at finding the perfect language and notation to conjure up mathematical ideas in others, that it's hard to think of him as aspergic. Yet, with him it's one-way traffic – Conway the performer. He doesn't seem remotely interested in what anyone else has to say unless it's about hardcore mathematics. It's almost as though he is compelled to keep up a constant flow of stories and anecdotes and ideas to forestall any possibility of normal two-way human interaction. He admitted once in an interview:

> I have a very odd sort of memory. I can remember the most useless, obscure details but when it comes to things that other people think important, I can't recall them for the life of me. When I was at Cambridge, I never learnt the name of some of my colleagues – and I worked with them for twenty years!

McKay wandered over to see what everyone was up to. 'Do you know how moonshine is translated into Chinese?' The word moonshine is quite difficult to translate into other languages because it can lose the nuances it has in English. 'Apparently they use the name for a man who sells the plot of land where he grows his rice, the very thing his family relies on.' He spotted the Japanese mathematician Harada on the other side of the room. 'Harada, Harada . . . how do you translate moonshine?' It turned out that Japanese mathematicians simply use the word for the light that shines off the moon. McKay was a bit disappointed that his story had fallen flat. 'I think moonshine is a bad name anyway. It put off the heavyweights who might have gone into the subject.'

Norton proudly spilled the content of his wallet onto the table to show us the bus and train tickets he had collected over the past few months. He pulled out one card he was particularly proud of: his membership of the Harwich, Felixstowe & Shotley Foot Ferry Society. He was also keen to give us details of his journey to the conference. While the rest of us had found the most direct route from our home to Edinburgh, Norton's extraordinary itinerary included several connections at small, provincial stations which showed off his grasp of the nation's train network. 'Are you going back the same way?' someone asked innocently. 'I don't think so!' Norton replied incredulously. He'd planned his trip back to take advantage of a boat crossing that only operates on a Wednesday.

Although the conference had another week to run, this was my last night. I had stolen a week away from my real family to celebrate 25 years of moonshine. But I didn't want to be away for too long. We'd spent the previous seven months living in Guatemala. Our plan to adopt children had led us to an orphanage in Antigua, where we met the identical twin girls who were to become part of our family. Shani and I decided to adopt them earlier that year, although the decision had nothing to do with symmetry. We'd just arrived home, a bigger family, when I went up to Scotland for the conference.

July two years on and the twins, Magaly and Ina, are three years old and starting to count. Although genetically identical they are very different; life isn't always as clear cut as mathematics. Next month the four-year cycle of the ICM brings another international gathering of mathematicians. Another round of Fields Medals will be awarded to the Borcherds of the day, but I am four months too old and will have to content myself with looking on from the sidelines. I've got a runner's-up prize. It is considered an honour to receive an invitation to present your work at the Congress, and I am preparing my lecture for next month. Last year my birthday was spent sitting on a beach in Sinai. This year on my birthday I've got to give a lecture at four in the afternoon.

Looking back over the year, the problem I have been working on has probably ended up getting more complicated than it was 12 months ago. It will make the final resolution, if it comes, that more gratifying. What's the satisfaction in solving easy problems? I'm still not sure even what the final answer will be.

Borcherds is right. In mathematics the real prize is not a medal or an invitation to the ICM, but making the breakthrough on the problem you've dedicated your life to. The prize might be claimed at any time and any place: on a broken-down bus in Kashmir, on a Saturday in Cambridge at twenty past midnight, or while listening to the engaged signal on the end of a telephone line in Bonn.

Further Reading

Many of the following books and papers provided material which was important in enabling me to write this book. For those who have been stimulated to dig deeper into the subject, I recommend reading any of them. I have decided not to list any of the highly technical material that requires a mathematics degree to appreciate, unless it contains some interesting non-technical insights.

Artmann, B. 'Roman dodecahedron'. *The Mathematical Intelligencer*, vol. 15 no. 2 (1993), pp. 52–53.

Artmann, B. 'A Roman icosahedron discovered'. *The American Mathematical Monthly*, vol. 103 no. 2 (Feb. 1996), pp. 132–133.

Aschbacher, M. 'The status of the classification of the finite simple groups'. *Notices of the American Mathematical Society*, vol. 51 no. 7 (Aug. 2004), pp. 736–740.

Aschbacher, M. 'Highly complex proofs and implications of such proofs'. *Philosophical Transactions of the Royal Society A*, vol. 363 no. 1835 (2005), pp. 2401–2406.

Baas, N. A. 'Sophus Lie'. *The Mathematical Intelligencer*, vol. 16 no. 1 (1994), pp. 16–28.

Baron-Cohen, S. *The Essential Difference. Men, Women and the Extreme Male Brain* (London: Allen Lane, 2003).

Borcherds, R. E. 'What is . . . the Monster?' *Notices of the American Mathematical Society*, vol. 49 no. 9 (Oct. 2002), pp. 1076–1077.

Borges, Jorge Luis. *Labyrinths* (London: Penguin, 1970).

Boyer, Carl B. *A History of Mathematics*, 2nd edn, revised by Uta C. Merzbach (New York: Wiley, 1989).

Cameron, P. J. and van Lint, J. H. *Designs, Graphs, Codes and their*

Links, London Mathematical Society Student Texts, 22 (Cambridge University Press, 1991).

Conway, J. H. 'Monsters and moonshine'. *The Mathematical Intelligencer*, vol. 2 no. 4 (1980), pp. 165–171.

Conway, J. H., Curtis, R. T., Norton, S. P., Parker, R. A. and Wilson, R. A. *Atlas of Finite Groups* (Oxford: Clarendon Press, 1985).

Conway, J. H. and Huson, D. H. 'The orbifold notation for two-dimensional groups'. *Structural Chemistry*, vol. 13 nos. 3/4 (Aug. 2002), pp. 247–257.

Conway, J. H. and Sloane, N. J. A. *Sphere Packings, Lattices and Groups*, 3rd edn. Grundlehren der mathematischen Wissenschaften, 290 (Berlin: Springer-Verlag, 1999).

Coxeter, H. S. M. *Regular Polytopes* (New York: Dover, 1973).

Cromwell, Peter R. *Polyhedra* (Cambridge University Press, 1997).

Davies, B. 'Whither mathematics?' *Notices of the American Mathematical Society*, vol. 52 no. 11 (Dec. 2005), pp. 1350–1356.

Fauvel, John, Flood, Raymond and Wilson, Robin (editors). *Music and Mathematics: From Pythagoras to Fractals* (Oxford University Press, 2003).

Fauvel, John and Gray, Jeremy (editors). *The History of Mathematics – A Reader* (London: Palgrave Macmillan, 1987).

Fritzsche, B. 'Sophus Lie. A sketch of his life and work'. *Journal of Lie Theory*, vol. 9 (1999), pp. 1–38.

Gray, J. 'Otto Hölder and group theory'. *The Mathematical Intelligencer*, vol. 16 no. 3 (1994), pp. 59–61.

Grünbaum, B. 'What symmetry groups are present in the Alhambra'. *Notices of the American Mathematical Society*, vol. 53 no. 6 (June/July 2006), pp. 670–673.

Grünbaum, B., Grünbaum, Z. and Shephard, G. C. 'Symmetry in Moorish and other ornaments'. *Computers and Mathematics with Applications*, vol. 12B nos. 3/4 (1986), pp. 641–653.

Grünbaum, B. and Shephard, G. C. *Tilings and Patterns* (San Francisco: W. H. Freeman, 1987).

Guy, R. K. 'John Horton Conway', in D. J. Albers and G. L. Alexanderson (editors), *Mathematical People: Profiles and Interviews* (Boston: Birkhäuser, 1985), pp. 43–50.

Hardy, G. H. *A Mathematician's Apology* (Cambridge University Press, 1940).

Hargittai, I. 'John Conway – Mathematician of symmetry and everything else'. *The Mathematical Intelligencer*, vol. 23 no. 2 (2001), pp. 6–14.

Hargittai, I. and Hargittai, M. *Symmetry: A Unifying Concept* (Bolinas, CA: Shelter Publications, 1994).

Hargittai, I. and Laurent, T. C. (editors). *Symmetry 2000*, Parts 1 and 2 (London: Portland Press, 2002).

Hawkins, T. 'The birth of Lie's theory of groups'. *The Mathematical Intelligencer*, vol. 16 no. 2 (1994), pp. 6–17.

Humphreys, J. F. and Prest, M. Y. *Numbers, Groups and Codes* (Cambridge University Press, 2004).

James, Iaon. 'Autism in mathematicians'. *The Mathematical Intelligencer*, vol. 25 no. 4 (2003), pp. 62–65.

Joyner, David. *Adventures in Group Theory* (Baltimore: Johns Hopkins University Press, 2002).

Land, Frank. *The Language of Mathematics* (London: John Murray, 1960).

Martínez, Guillermo. *The Oxford Murders* (London: Abacus, 2005).

Matossian, Nouritza. *Xenakis* (Nicosia: Moufflon Publications, 2005).

Matte Blanco, Ignacio. *The Unconscious as Infinite Sets* (London: Karnac, Maresfield Library, 1998).

Neumann, P. M. 'On the date of Cauchy's contribution to the founding of the theory of groups'. *Bulletin of the Australian Mathematical Society*, vol. 40 (1989), pp. 293–302.

Neumann, P. M. 'A hundred years of finite group theory'. *The Mathematical Gazette*, vol. 80 no. 487 (Mar. 1996, centenary issue), pp. 106–118.

Neumann, P. M. 'What groups were: A study of the development of the axiomatics of group theory'. *Bulletin of the Australian Mathematical Society*, vol. 60 (1999), pp. 285–301.

Pérez-Gomez, R. 'The four regular mosaics missing in the Alhambra'. *Computers and Mathematics with Applications*, vol. 14 no. 2 (1987), pp. 133–137.

Schattschneider, Doris. *M. C. Escher: Visions of Symmetry* (London: Thames & Hudson, 2004).

Schattschneider, D. and Emmer, M. (editors). *M. C. Escher's Legacy: A Centennial Celebration* (New York: Springer, 2003).

Seife, Charles. 'Mathemagician (impressions of Conway)'. *The Sciences* (May/June 1994), pp. 12–15.

Singh, Simon. 'Interview with Richard Borcherds'. *Guardian*, 28 August 1998.

Solomon, R. 'On finite simple groups and their classification'. *Notices of the American Mathematical Society*, vol. 42 no. 2 (Feb. 1995), pp. 231–239.

Solomon, R. 'A brief history of the classification of the finite simple groups'. *Bulletin of the American Mathematical Society* (new series), vol. 38 no. 3 (2001), pp. 315–352.

Stewart, Ian. *Galois Theory*, 3rd edn (London: Chapman & Hall, 2004).

Struik, Dirk J. *A Concise History of Mathematics*, 4th revised edn (New York: Dover, 1987).

Stubhaug, Arild. *Niels Henrik Abel and His Times* (Berlin: Springer-Verlag, 2000).

Stubhaug, Arild. *The Mathematician Sophus Lie* (Berlin: Springer-Verlag, 2002).

Thompson, Thomas M. *From Error-Correcting Codes Through Sphere Packing to Simple Groups*. The Carus Mathematical Monographs, no. 21 (Washington: Mathematical Association of America, 1983).

Tignol, Jean-Pierre. *Galois' Theory of Algebraic Equations* (Singapore: World Scientific, 2001).

Toti Rigatelli, Laura. *Évariste Galois: (1811–1832)*, Vita mathematica, vol. 11 (Boston: Birkhäuser, 1996).

Welsh, Dominic. *Codes and Cryptography* (Oxford University Press, 1988).

Weyl, Hermann. *Symmetry* (Princeton University Press, 1952).

Xenakis, Iannis. *Formalized Music: Thought and Mathematics in Music* (New York: Pendragon Press, 1992).

Websites

http://www.maths.ox.ac.uk/~dusautoy My homepage contains a selection of archived material both from mathematical journals and the mainstream media.

http://www-groups.dcs.st-and.ac.uk/~history/ A wonderful source of mathematical biographies maintained by the University of St Andrews.

http://mathworld.wolfram.com A good site for accessing more technical definitions and explanations of the mathematical material.
http://findingmoonshine.blogspot.com/ My blog which records the
ongoing saga of my search for symmetry and extra resources relevant
to the book.

All the articles in the above list from the *Notices of the American
Mathematical Society* and the *Bulletin of the American Mathematical
Society* are available on line at http://www.ams.org/notices/ and http://
www.ams.org/bull/

Acknowledgements

I should like to thank my fellow mathematicians who generously gave me their time and stories during the writing of this book. Fourth Estate and HarperCollins have been wonderful publishers, full of people who really care about books. My editor Mitzi Angel was an invaluable partner in charting a course through a complicated narrative. My copyeditor John Woodruff's attention to detail, both mathematical and linguistic, is gratefully acknowledged. Antony Topping of Greene and Heaton is more like my personal psychologist than an agent and was there by my side throughout the writing process. Wadham College and the University of Oxford have done an immense amount to facilitate my research and to support my efforts to bring mathematics alive to those beyond the ivory towers. My research council, the EPSRC, deserve especial thanks for creating the Senior Media Fellowship that gave me the space over the last few years to be a mathematical ambassador as well as a mathematician. Biggest thanks are due to my family, Shani, Tomer, Magaly and Ina, who joined me on my journey through symmetry.

Picture credits

1 Fibonacci's spiral © Raymond Turvey
2 Icosahedron © Raymond Turvey
3 Six-pointed starfish © Raymond Turvey
4 Swapping symmetries © Raymond Turvey
5 M.C. Escher's Tin for the Verblifa Co. © 2007 The M.C. Escher Company-Holland
6 Changing geometry into number © Raymond Turvey
7 Symmetries of a triangle and pentagon © Raymond Turvey
8 Icosahedron © Raymond Turvey
9 Diamond, graphite and buckyball © Raymond Turvey
10 Four-dimensional cube © Raymond Turvey
11 Groups of spirals at Newgrange tumulus in County Meath © Raymond Turvey
12 Kite © Raymond Turvey
13 Neolithic stone balls © Ashmolean Museum
14 Carved symmetrical ball found at Towie in Scotland © The Trustees of the National Museums of Scotland
15 Tetrahedral dice © Raymond Turvey
16 Combination lock © Raymond Turvey
17 Supermarket grid © Raymond Turvey
18 Pyramids © Raymond Turvey
19 Octahedron © Raymond Turvey
20 Octagons © Raymond Turvey
21 Icosahedron © Raymond Turvey
22 Icosahedron and sphere of 12 pentagons © Raymond Turvey
23 Symmetry in pavements © Marcus du Sautoy
24 Eight-pointed star © Marcus du Sautoy

Index

P.S.

Ideas,
interviews
& features . . .

Portrait of Marcus du Sautoy

by Roger Tagholm

VISITING MARCUS DU SAUTOY at home in Stoke Newington, north London, is a delightful experience. One minute he's talking about multi-dimensional space, the next his twin adopted five-year-old daughters are wobbling into his study with slices of cake made by his Israeli wife Shani. If mathematics can at times seem abstruse and far removed from everyday life, there is nothing like young children to bring you down to earth.

In a sense, this is what du Sautoy is all about: the marrying of high maths – the language of all the sciences – with everyday life. He believes, passionately, that 'science only comes alive when you communicate it to other people', and as such he is the perfect successor to Richard Dawkins for the prestigious Charles Simonyi Professorship for the Public Understanding of Science. Du Sautoy loves using ordinary subjects to illustrate his points, such as football.

'Talk Sport radio wanted to know if there are any equations Wayne Rooney is using when he plays football. There is a huge amount of maths behind playing football. When you take a free kick and want to work out where the ball will land, you have to solve a quadratic equation to do that. Now of course, Wayne Rooney is not solving quadratic equations when he takes a free kick, but he is doing that subconsciously.'

Which brings du Sautoy to symmetry's role in evolution. Just as those who can calculate the trajectory of a ball most accurately may win the football match, so those who could do the same for an arrow or cannonball would have found food or won the battle. 'We're evolutionarily programmed to be sensitive to symmetry because those that can pick out symmetry, who can start to read its messages, will survive. For example, the bumblebee. It has very bad vision, it sees the world in black and white, it has no sense of distance. But what it can pick out is shape and symmetry – and the symmetry of the flower is saying here is where sustenance is.'

His study tells you a lot about the man. As with many creative people, it isn't tidy. Papers and books swirl around his Apple laptop and are even on the floor under the desk. Two giant, brightly coloured, odd-shaped dice are on a top shelf above the piano, and his son's digital drum kit lies just below. Gazing down on it all is an old print of Pierre François du Sautoy, a Catholic Frenchman who helped Bonnie Prince Charlie fight the English in the eighteenth century and from whom du Sautoy is descended.

Colours seem to appeal to this top mathematician, perhaps as an antidote to the occasional dryness of his chosen subject. He favours purple trousers and today is wearing them with red-and-black square-patterned socks. It would be hard to describe a typical day for him. He'll do the school run, unless he's away, and then he might find himself whisked off to a photo shoot for *Esquire* magazine, or to the BBC to record *Desert* ▶

6 There is a huge amount of maths behind playing football. When you take a free kick and want to work out where the ball will land, you have to solve a quadratic equation to do that. 9

Author photo © Niall McDermid

LIFE
at a Glance

BORN

1965 in London

EDUCATION

Attended Gillotts School, a local comprehensive, then read Maths at the University of Oxford

CAREER

Currently Professor of Mathematics and a Fellow of New College, Oxford. Author of numerous academic articles and books on mathematics, and has also written and presented a number of TV programmes on the subject. Has been a visiting professor at the École Normale Supérieure in Paris, the Max Planck Institute in Bonn, the Hebrew University in Jerusalem and the Australian National University in Canberra. In 2008 he took up the Charles Simonyi Professorship for the Public Understanding of Science. ▶

Portrait of Marcus du Sautoy (continued)

◀ *Island Discs*, or taking time to scribble down some ideas for his 'Sexy Maths' column in *The Times*. He enjoys talking to schools, but says he wants to devote much of his energy now to using the media to reach even more people, as he did with the BBC4 series *The Story of Maths*.

Curiously, all the time this is happening, some part of his mind will be churning over a particular problem. Essentially, he is interested in calculating what possible symmetries there are, both in three dimensions and higher dimensions. 'Although the book has an end because we created a periodic table to symmetry, that's just a new beginning because now we need to know how we put those things together. What are the molecules you can make out of these atomic symmetries? That is a total mystery. Maths often feels like a Greek mythical beast. You lop off one head, and the discoveries you make through doing that mean that two more heads pop up that you've got to fight with, and that's what makes it an exciting subject. Each new breakthrough leads to new questions.'

With du Sautoy wielding the metaphorical sword, the maths creature has met a formidable – but immensely likeable – opposition. ∎

du Sautoy on …

House numbers

'We used to live at number 53, which is a prime number, and that number was on the cover of the hardback of *The Music of the Primes*. I was slightly disappointed when we moved to number 1. It used to be thought of as a prime because it can only be divided by 1 and itself. That's the technical definition. But the important thing about primes is that they are building blocks – you build new numbers from them. They're the atoms, the hydrogen and oxygen, but 1 is a little bit like the vacuum. If you multiply by 1 you get nothing new. So 1 has been cast out.'

Telephone numbers

'When we moved here I had to get a new telephone number and I thought, cool, I can get a prime number telephone number. But London telephone numbers are eight digits long, and Gauss's prime number theorem tells you that by then, the chances of that number being prime are 1 in 15. So when I rang and was given the new number I quickly put it into my computer and discovered it wasn't prime. I told the woman that I couldn't remember that number, so she gave me another. But that wasn't prime either and I think she was now getting so fed up that she gave me an even telephone number, of all things.'

And football

'Beckham crosses. A perfectly timed volley by Rooney. Goal!!!' But how did Rooney do it?

The footballer Wayne Rooney has had to ▶

LIFE *at a Glance*
(continued)

◀ PRIZES AND AWARDS

In 2001 he won the prestigious Berwick Prize of the London Mathematical Society, awarded every two years to reward the best mathematical research by a mathematician under forty.

FAMILY

Married with children

LIVES

Stoke Newington, London

du Sautoy on ... *(continued)*

◀ solve not one but three complicated equations to judge the flight of the ball as it comes flying into the box.

$$y = \left(\frac{v}{u}\right)x - \left(\frac{g}{2u^2}\right)x^2$$

$$F_D = C_D\rho KAv^2/2$$

$$y'' + akxy' + by' + cy = g$$

The first equation, discovered by the great Galileo, determines the height of the ball given its starting velocity when it leaves the foot of David Beckham. Obviously, Rooney needs to solve this quadratic equation to know where the ball is going to land, so that he can volley it before it touches the ground. Once you know the incoming ball's horizontal velocity u, and vertical velocity v, just set y to be the perfect height off the ground for your volley or header, and then solve for x.

But there are some complications he needs to factor in. The second equation controls the drag factor on the ball, which changes the ball's velocity as it flies through the air. The smooth surface of the ball has a strange effect, causing the drag to be very small at first, then suddenly, towards the end of its flight, the drag kicks in, slowing the ball dramatically. It is quite counter-intuitive and has caught out many a goalie.

Thirdly, what about that wicked spin that Beckham can put on the ball? That's controlled by the third equation – the boomerang equation – a second order differential equation.

Having worked out the flight of the ball, Rooney needs to solve the equations all over again to find the most efficient path from his foot to the goal whilst avoiding any obstacles such as a defender on the way. As the great Dennis Bergkamp once said: 'Every kick of the ball requires a thought.' ■

Einstein, Plato ... and you?

by Marcus du Sautoy

THE EINSTEINIUM ATOM, the Higgs boson, Halley's Comet. A new scientific discovery always provides the creator with the chance to choose a name for their finding. Sometimes it is their own name that they hope will stick, giving them the chance of a little bit of immortality. But recently, scientists have been honouring some of their inspirations and heroes.

Neil Young found his name on a new species of trapdoor spider, the *Myrmekiaphila neilyoungi*, discovered last year by a biologist who was a fan. David Attenborough has been immortalized in numerous animals although he seems to specialize in extinct creatures, including the *Attenborosaurus* and more recently the *Materpiscis attenboroughi*, a fossil demonstrating the earliest example of a fish that gives birth to its young rather than spawning eggs. Even politicians have made it onto the roll call: George W. Bush gave his name recently to a new species ... of slime-mold beetle.

There has always been a tradition of honouring great pioneers by naming an object after them. The craters on the moon read like a Who's Who of historical characters from Copernicus to Plato, Archimedes to Galileo. The features on the moons of Venus are reserved for female heroines: Agatha Christie, Anne Frank and Barbara Hepworth have all been recognized. The rocks tumbling around in the asteroid belt have also honoured a slew of famous names – Newtonia, Mozartia, Gaussia after the mathematician who charted the path of the first asteroid Ceres.

But even if you're not as famous as Newton or Neil Young, do not despair. There is a whole industry out there providing individuals with the chance to get their own names on stars, although the central registry kept by the International Astronomical Union prefers numerical nomenclature to navigate their way round the night sky, rather than referring to the Auntie Mabel nebula. There are even companies willing to sell individuals plots of land on Mars or the moon after it was discovered there was a loophole in the United Nations Outer Space Treaty of 1967 which outlawed countries or governments laying claim to extraterrestrial land but forgot to include people. But if outer space isn't your thing how about having your name on a mathematical object in hyperspace instead? Well, now is your chance.

Mathematicians are constantly on the lookout with their telescopic sights for new symmetrical objects out there in the mathematical firmament. The first symmetrical objects to be discovered were dice and shapes like the football made out of patches of hexagons and pentagons. But as mathematics became more sophisticated mathematicians started concocting symmetrical shapes beyond the three dimensions we live in. These shapes aren't simply good for games but turn out to ▶

6 During the cold war Soviet and Western mathematicians often came to blows at conferences over the differing names that were attached to ideas discovered simultaneously in the East and West. 9

The Alhambra, Granada, Spain: The tiling seen in this fabulous Muslim palace explores the seventeen different symmetries that are possible on a two-dimensional surface.

Arche de la Défense, Paris: Intriguing because this is the *shadow* of a four-dimensional cube. The symmetries of such a shape are related to symmetries discovered by the Norwegian mathematician Sophus Lie.

Game of Ur, British Museum, London: The little tetrahedral-shaped dice used in this ancient game were the first symmetrical objects used by humans.

Nikko, Japan: All the symmetries on columns supporting a gate on one of the site's Shinto shrines are identical except for one. The Japanese believe that leaving something unfinished makes the whole more interesting.

Einstein, Plato … and you? *(continued)*

◄ explain the behaviour of viruses and crystals, to be at the heart of many of the codes used in modern technology, and even to explain the menagerie of fundamental particles that make up matter itself.

As soon as one of these new symmetrical objects is spotted it needs a name. Just as with children, the naming of a mathematical object is part of giving birth to your creation. It is what gives it its own identity, distinguishing it from all the other mathematical objects out there. And just as children provide the parents with the hope of continuing their genetic inheritance, so mathematical creations provide the architect with a chance of immortality.

But the naming of mathematical objects is fraught with difficulties. Unlike in astronomy, there is no central registry where you can record your choice of name. It is only by a process of communal acceptance and use that a name takes off. The rather unregulated attitude to mathematical nomenclature has of course led to some bitter rivalries. During the cold war Soviet and Western mathematicians often came to blows at conferences over the differing names that were attached to ideas discovered simultaneously in the East and West.

In the story of symmetry, the mechanism by which new objects came to light often fuelled passions over who should really get their name on the shape. There has often been a two-stage process on the way to a new discovery. First a mathematician would come up with evidence for the possible existence of a mathematical object with a given number of symmetries. The evidence would then inspire other mathematicians to actually create the new object. But whose name should be given to the object? Frequently it was the name of the person who predicted it who got immortalized, much to the chagrin of the person who put in the hard graft of crafting the symmetrical shape.

The same process happens in physics. Fundamental particles are often predicted before they are ever seen for the first time in the particle accelerator. The theoretical analysis guides the experimenters to know what they should be looking for. The Higgs boson has never been observed but already it has a name after Peter Higgs suggested that its existence would solve some of the mysteries of matter. Interestingly it was the mathematics of symmetry which led to this prediction. The hope is that with the Large Hadron Collider at CERN we might see the particle for the first time. However, the person who first sees it has already missed their chance to get their name on the particle, although there is great hope that a whole host of new particles might still be up for grabs.

Sometimes new symmetrical objects get more exotic names. In the early 1970s the German mathematician Bernd Fischer predicted the existence of an extraordinary symmetrical object with more symmetries than there are atoms in the sun, which only appears when you enter 196,883-dimensional space. It was totally unrelated to anything else anybody had seen before. Its terrifying proportions led the Cambridge ▶

The Symmetry
Tour – *continued*

National Museum of Football, Preston: The way footballs have developed over the last century demonstrates the different symmetrical shapes you can use. The ball used for the World Cup in Germany in 2006 used eye-patch-type shapes and triangles. There's an Archimedean solid hiding behind there which is fundamentally different from the usual pentagons and hexagons.

Einstein, Plato … and you? *(continued)*

◄ mathematician John Conway to christen the group simply The Monster.

Some mathematicians doubted if it was possible to conceive of such a ridiculously complex symmetrical object. So when in 1980, the American mathematician Bob Griess finally gave flesh to this Monster, he understandably wanted his name somehow attached to this extraordinary creation. He spearheaded a drive to have the group renamed. He realized he couldn't hope to convince the mathematical community to give up the colourful name of the Monster simply for the Fischer-Griess group. So he went instead for renaming it The Friendly Giant. To see the F and the G there to recognize his and Fischer's involvement would suffice. The name never took off.

My own mathematical explorations have revealed a whole seam of interesting symmetrical objects that connect with one of the major themes of modern mathematics: elliptic curves. These equations like $Y^2 = X^3 - X$ are some of the most fascinating in mathematics and were key to the resolution of Fermat's Last Theorem. It is one of mathematics' holy grails to understand what choices of whole numbers X and Y will solve equations like this one. Indeed one of the Clay problems for which one can win a million dollars concerns understanding which elliptic curves have solutions and which do not.

But these new symmetrical objects are in need of names. So I thought if you can give someone a star for their birthday, celebrate an anniversary with territory on Mars, or even buy a pixel in a picture for a dollar, why not let people adopt a symmetrical object? I support a charity called Common Hope that helps street kids in Guatemala get a decent education, health care and shelter. In exchange for a donation to the charity, contributors will get one of my new symmetrical shapes named after whichever hero or loved one they would like honoured.

I can't promise that your symmetrical object will explain the particles in the LHC that make up the fabric of the universe, or will be the key to creating a new code, or even will help solve these million-dollar equations. But you never know. Even if they just turn out to be useless symmetrical playthings of the mind, at least they will have helped young children's lives in one of the most impoverished countries in the world. Who said maths couldn't save the world?

To name a symmetrical shape, make a contribution at www.firstgiving.com/findingmoonshine
The shapes are exhibited on the blog which accompanies this book: www.findingmoonshine.blogspot.com

A version of this article first appeared in the *Daily Telegraph* on 15 July 2008 ■

❝ If you can give someone a star for their birthday, celebrate an anniversary with territory on Mars, or even buy a pixel in a picture for a dollar, why not let people adopt a symmetrical object? ❞

Have You Read?

Other books by Marcus du Sautoy

Music of the Primes

At school, children are taught that prime numbers are divisible only by themselves and the number one. What they are not taught is that primes represent the most tantalising enigma in the pursuit of human knowledge. In his first book Marcus du Sautoy opens up the wonderful world of prime numbers and reveals some of the secrets it contains.

The Num8er My5teries

Every time we download a song from i-tunes, take a flight across the Atlantic or talk on our mobile phones, we are relying on great mathematical inventions. Maths may fail to provide answers to various of its own problems, but it can explain some of the everyday mysteries in the world around us – how prime numbers are the key to Real Madrid's success, to secrets on the Internet and to the survival of insects in the forests of North America. ∎

If You Loved This,
You Might Like ...

The Symmetries of Things
John H. Conway, Heidi Burgiel, Chaim Goodman-Strauss
Hearing John Conway talking about symmetry is always a privilege. Here with his co-authors he gives a tough but engaging account of what symmetry is. A great book if you would like to learn more about the symmetries underlying the Alhambra.

Taming the Infinite: The Story of Mathematics
Ian Stewart
Ian Stewart provides a wonderful tour through the history of mathematics from Ancient Babylon through to the modern day.

Poincaré's Prize: The Hundred-Year Quest to Solve One of Math's Greatest Puzzles
George G. Szpiro
One of the most exciting stories of the last century is told by George Szpiro as he explains how the Russian recluse Grigori Perelman finally solved the Poincaré conjecture.

FIND OUT MORE

www.findingmoonshine. blogspot.com
The website for this book contains a regularly updated news section, Marcus du Sautoy's online journal and a huge amount of fascinating information related to symmetry.

www.maths.ox.ac.uk/ ~dusautoy
Marcus du Sautoy's homepage contains a selection of archived material both from mathematical journals and the mainstream media.

Three Number Puzzles

Magic seven (grade: easy)
Think of a number from 1 to 10. Double it. Add 14. Divide by 2. Take away the number you originally thought of. The number you are left with is 7.

Fiendish eleven (grade: hard)
Take any four-digit number. For example: 1234. Make 1 = A, B = 2, etc. Now write the same number as BCDA, i.e. 2341. Add both numbers.

1234 + 2341 = 3575. The final figure will always be divisible by 11. This will work with six-digit numbers, eight-digit numbers, etc., but *not* with odd-digit numbers.

And fun with ISBNs
Divisibility by 10 is the key to a code used by publishers. Take the ISBN for this book which you can find on the back. It has 13 digits. The first 12 digits provide information about the publisher, author and title. The 13th digit is added as a check digit to pick up mistakes. How does this work? Add up the 2nd, 4th, 6th, 8th, 10th and 12th digit and multiply the sum by 3. Now add on the other digits. The total will be divisible by 10. If you make a mistake in writing down the ISBN, then often the calculation will give you a number not divisible by 10. The calculation behind the ISBN is similar to the symmetrical codes used to correct errors in digital data. ∎